Beachcomber's Guide to CALIFORNIA MARINE LIFE

Thomas M. Niesen

Photography by Michael E. Kunz

Line Drawings by David I. Wood

Gulf Publishing Company
Houston, London, Paris, Zurich, Tokyo

> To all my friends and students whose curiosity, interest, enthusiasm, and encouragement have kept me young and the juices flowing; and to my family, Anne and Amy, who make it worthwhile.

Gulf Publishing Company
Book Division
P.O. Box 2608 • Houston, Texas 77252-2608

10 9 8 7 6 5 4 3 2 1

Printed in the United States of America

Library of Congress Cataloging-in-Publication Data
Niesen, Thomas M.
  Beachcomber's guide to California marine life / by Thomas M. Niesen;
black & white photography by Michael E. Kunz; line drawings by David I. Wood;
color photography by author.
    p. cm.
  Includes bibliographical references.
  ISBN 0-88415-075-5
  1. Marine biology — California. 2. Seashore biology — California.
  3. Beachcombing — California — Pacific Coast.   I. Title.
QH105.C2N54 1994
574.9794—dc20

                                               93-29207
                                                 CIP

Contents

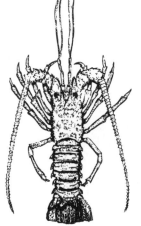

CHAPTER 7   **The Rocky Intertidal Environment:
Middle and Low Zones**   **141**

CHAPTER 8   **Marine Mammals of California**   **174**

Acknowledgments

Over the years I have had the benefit of association with some of the West Coast's finest marine biologists. At the University of California at Santa Barbara I studied with Drs. Demorest Davenport and Alfred Ebeling. During my brief stay at Humboldt State University I came to know Drs. John DeMartini and Jack Pearce. At San Diego State University I was able to study with Drs. Richard Ford, William Hazen, Norman McLean, and Deborah Dexter. At the University of Oregon I benefited greatly from the help of Drs. Peter Frank, Paul Rudy, Robert Terwilliger, and Bayard McConnaughey. As a faculty member at San Francisco State University over the last 20 years, I have had the privilege to work with Drs. Alissa Arp, Robert Beeman, Margaret Bradbury, Michael Josselyn, Hal Markowitz, Jack Tomlinson, Albert Towle, Hideo Yonenaka, and my diving partner Dr. Ralph Larson. Finally, I have been fortunate enough to work with Drs. Gregor Cailliet, Michael Foster, and James Nybakken at Moss Landing Marine Laboratories, and my cherished colleague, Dr. Edwin Lyke of California State University, Hayward. All of these scientists have shared their knowledge and love of marine biology with me and I hope I don't embarrass them by acknowledging their influence and importance to me.

More directly related to this *Beachcomber's Guide* I want to single out Michael Kunz for his extraordinary patience and consummate skill in black and white photography. Mike and I spent countless hours in the field armed with our cameras and strobes, and the black and white photos within are mainly his with an occasional "lucky" shot of my own. I would also like to thank David Wood for his expertise as an illustrator and biologist and Linda Johnson for all the logistic support. The many line drawings David prepared for this text, mainly used in Chapters 3, 4, 5, and 8 are exceptional. My dear friend and long-time collaborator, Suzanne Offutt Pouty, provided the line drawings of the sea urchin and encrusting bryozoan used in Chapter 3. Cheryl Aday used her computer wizardry to create the illustrations of *Cycloxanthops novemdentatus* and *Taliepus nuttallii* used in Chapter 7.

Monique Fountain, graduate student and sea-water technician extraordinaire, collected and maintained many of the animals brought into the lab for photographing. Monique also prepared the technical figures used in Chapters 1-4, and provided access to and valuable consultation on her shell collection from southern California. Another (former) graduate student and good friend, Paul Miller, lent me his underwater colored photograph of *Flabellinopsis iodinea* which is found in the color plate section of this text. I am very grateful for his generosity. I would also like to thank John (Mike) McFarland for sharing his vast knowledge of sandy beaches with me.

Dr. Ned Lyke provided carte blanche access and borrowing privileges to his excellent dry shell collection, which was the source of many of the shell illustrations used in the text. He also reviewed my initial outline for this guide and told me I was crazy to attempt it. I should have listened.

My wife, Anne Niesen, patiently read and corrected innumerable drafts of the text, always with good humor, excellent suggestions and much appreciated encouragement. My daughter, Amy, had the good sense to go away to college. Thank you both.

Finally, I'd like to thank the especially agreeable and unflappable Texas gentleman, Mr. Timothy Calk, my editor from Gulf Publishing. It was a pleasure working with such a professional throughout the odyssey of this text.

Tom Niesen, Ph.D.
Half Moon Bay, California

Foreword

Like the plants and animals described herein, *Beachcomber's Guide to California Marine Life* is well adapted to an unfilled niche. More comprehensive than Joel Hedgpeth's *Introduction to Seashore Life*, but less technical and better illustrated than *Light's Manual* or Rickett's classic *Between Pacific Tides*, this guide will surely benefit amateur intertidal naturalists and marine biologists. I have observed Dr. Niesen at work in his university classroom, a patient pedagogue but demanding scholar, and followed in his wet footsteps as he sought out to explore and explain the tidepool plants and animals that are his passion.

This guide may at first stagger the reader with such scientific jawbreakers as *Blepharipoda occidentalis* and *Cycloxanthops novemdentatus*, but fortunately, photographs and common names like mole crab and pebble crab are included. However, don't dismiss the scientific nomenclature that exists throughout, for learning sonorous sobriquets such as *Upogebia* (a mud shrimp) and *Busycotypus* (a whelk) is its own reward. Equally rewarding is this book's organization by habitat groupings, explanatory text of marine biology, geology, and chemistry, and the useful advice given concerning such diverse topics as dinoflagellate toxins (more accurate than recalling a month with an "r") and tidepool wear (viz., bring a change of clothes and plan on falling in).

As enjoyable as are the text, references, photographs, and illustrations of this fine guide, its greatest value is the conservation ethic it provides. Not by preaching but rather by the knowledge that it imparts will this guide convince the beachcomber to leave intertidal life in place for others to see, understand, and enjoy.

John E. McCosker, Ph.D.
Director, Steinhart Aquarium
San Francisco

Preface

❧

It has been my good fortune to know what I wanted "to be" from the age of 16. Growing up in southern California and blessed with a mother that loved the seashore, I was introduced to "the beach" at an early age. My mother would pack myself and my three brothers into her '48 Ford and haul us the ten miles to Cabrillo Beach in San Pedro, Los Angeles County. As I progressed from wading to body surfing to a surfboard, I likewise became more and more aware of and intrigued by the marine life that surrounded me. My moment of career revelation came one summer day during an especially strong south swell. The clear, warm water sweeping up from the south not only produced outstanding waves, it also harbored an abundance of marine life including a pack of thresher sharks. The sharks were clearly visible through the water beneath my surfboard, busily chasing one another in courtship. The idea struck me that I could continue to indulge my fascination with the sea and get paid for it!

Armed with my father's Germanic work ethic and access to California's excellent system of higher education, I was able to pursue my career in marine biology. Now after 30 years of learning and teaching marine biology I hope to share with you the sense of awe and wonder that I still feel beachcombing in California.

This book occupied my existence for the better part of two years. I still wake up thinking of things that I should have included or changed. I have tried to give beachcombers my best shot at covering the most common organisms without drowning them with trivia. I hope that at some level you feel I have succeeded. If this book helps you enjoy California's incredible coast, I know that I have. See you on the beach.

1
Introducing the California Coast

THE CALIFORNIA COASTLINE

Anyone exploring California's long coastline quickly becomes aware of its beauty and vast diversity. Beyond the next curve in the coastal highway may lurk a long sandy beach, a towering coastal headland, or a picturesque cove (Figure 1-1). At first these habitats may appear to pop up randomly along the coastline. However, there is a pattern to their distribution, and a brief consideration of the geology of California is helpful in understanding this pattern.

California sits along the eastern side of the Pacific Ocean and forms part of what is referred to as the Pacific Rim. The western side of the Pacific Rim consists of the Asian mainland and the Japanese and Philippine islands. The Pacific Rim is also called the Ring of Fire because of its numerous active volca

noes. The volcanoes along with the infamous California earthquakes are all manifestations of very active geological processes that are characteristic of the Pacific Rim.

The surface of our earth is not a continuous sheet, but instead is divided into a series of crustal plates. These plates float on the earth's molten core like a cracked but intact egg shell would sit on the surface of a hard boiled egg. These plates are not static, but are in motion. At sites on the deep seafloor where two oceanic plates join, molten rock (magma) wells up from below and adds new material to the edges of the plates. This causes the seafloor to spread and pushes the oceanic plates against the plates that make up the continents.

California is on the edge of a continental plate that butts up against an oceanic plate. As the seafloor spreads in mid-Pacific, the oceanic plate pushes

Figure 1-1. The dramatic California coastline. Towering bluffs frame Tunitas Creek (foreground, right) on the San Mateo County coast.

against the continental plate and slides along and underneath it. However, this slide is not continuous and smooth, rather it occurs in abrupt jolts that we know as earthquakes. As the two plates push against one another they temporarily stick along their margins. Over time pressure builds up that finally overcomes this resistance and the two plates slide. The numerous fault lines in California are locations where this active sliding has been mapped. Over the millions of years of geological time this constant pressure of the oceanic plate against the continental plate has folded and wrinkled much of the coastline of California into the low lying coastal hills and mountain ranges we see today. This seismic activity coupled with the rise and fall of sea level over the millennia have produced a dramatic and varied coastline available for the beachcomber to explore.

In the more recent past the California coastline has been influenced by the last ice age. About 10,000 years ago much of the northern hemisphere was covered by a thick ice sheet. The water contained in the ice sheet caused sea level to drop 450 feet. This exposed the entire shallow continental shelf along California's coast. There are places where the shoreline was over 30 miles west of its present location! As the ice sheet melted and sea level rose, many changes were wrought along California's coast. The advancing water weathered soft sedimentary rocks, cutting flat marine terraces and providing some of the sediment that formed many of our sandy beaches. Harder basaltic rocks that had been formed by volcanic processes became coastal headlands such as Cape Mendocino and Point Conception. Large river deltas that had formed on the uncovered continental shelf were submerged, and their channels remained as submarine canyons such as those found just offshore of Monterey Bay and La Jolla. Coastal river valleys were flooded with sea water and formed estuaries such as San Francisco Bay.

The coast north of Point Conception is relatively straight, indented by the major embayments of Monterey Bay and San Francisco Bay (Figure 1-2). The coastline here has many sheer cliffs that are weathered by waves sweeping in from the Pacific unabated to unleash their energy against the coast. South of Point Conception, the coastline takes a decidedly eastward turn, and the fringing Channel Island archipelago serves to break the back of the Pacific waves before they reach the relatively sheltered coast. The coastline south of Point Conception consists of low coastal sea cliffs and a few marine terraces.

PHYSICAL FACTORS INFLUENCING MARINE ORGANISMS

The domain of the beachcomber is that area where the ocean meets the land. This area is known as the intertidal zone or littoral zone. Wave action is one of three major physical factors that influence what the beachcomber will discover in any given marine habitat. The other two factors are the type of bottom substrate and the degree of tidal exposure to air. The amount of wave exposure a given habitat receives is due to a combination of the proximity of offshore protection, the direction of the prevailing waves, and the geographic orientation of the habitat. For example, north of Point Conception there are no islands and few offshore submarine banks to intercept waves. Waves approach the shoreline primarily from the northwest. Therefore, a rocky headland facing to the northwest will experience the full brunt of the heavy Pacific waves. The only organisms capable of living here will have to be able to attach and grow under the onslaught of the pounding waves. Behind the headland, a small sheltered cove that faces south will experience decidedly less wave action and many, more fragile organisms may survive and flourish here.

Rocky substrate habitats provide many places for organisms to live. Depending on the nature of the rock and how it weathers, rocky areas can have considerable topographic relief. Broad rocky reefs with ledges, crevices, outcroppings, boulder fields and tidepools provide myriad niches for organisms to occupy. Soft substrates, such as the sand on a sandy beach or the mud of an estuarine mud flat, provide many fewer sites for organisms. Because the surface of a soft substrate provides no fixed sites for attachment and is usually quite flat and exposed, the only place to live is beneath the substrate surface. As a general rule, the marine community living in a soft substrate habitat has a relatively low diversity of organisms, but can have very high numbers of individual species. Hard substrate communities tend to be highly diverse with smaller numbers of individual organisms.

The final physical factor that influences what a beachcomber will encounter is the degree of tidal exposure a habitat receives. The higher an organism is located relative to the low tide mark, the longer it will be exposed to air. Most marine organisms must remain moist to breathe, and exposure to air dries them out. Not surprisingly, the only organisms a beachcomber will find in the highest tidal zone will be those that have special adaptations or behaviors to avoid drying out. As you move lower into the intertidal zone, the degree of exposure an organism encounters is reduced and more diversity will be discovered, with the most diverse areas occurring at tidal

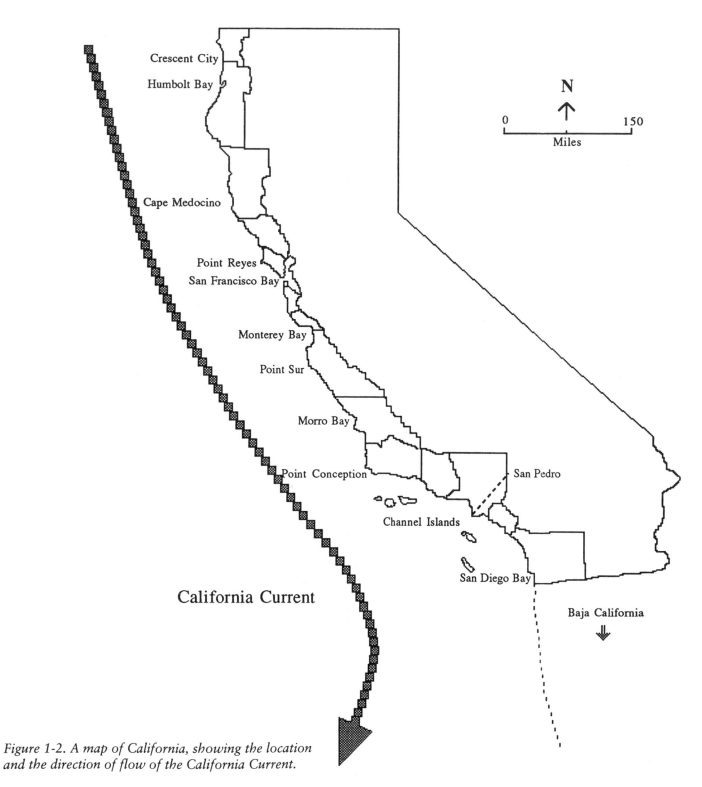

Figure 1-2. A map of California, showing the location and the direction of flow of the California Current.

levels exposed by only the lowest tides. This effect is especially noticeable in rocky substrate habitats where organisms are attached to the rocks and directly exposed to air. The effect of tidal position can be somewhat reduced by organisms living in soft substrate habitats if the substrate remains saturated with water at low tide.

WHY IS THE WATER SO COLD?

A recurring summer scene on California beaches is the neophyte beachcomber dashing into the surf only to beat a hasty retreat right back out again. This scene is especially predictable if the beach is in central or northern California, and the beachcomber has spent some time on the East Coast in the summer time. Why is the water so cold?

There are several reasons for this. First of all California is on the east side of the Pacific Ocean. In the ocean basins of the northern hemisphere surface water circulates in a large clockwise eddy or spiral. In the Pacific the water warms at the equator, flows northward along the coastline of Asia, and west in the chilly northern latitudes. When the water begins its southward flow along our coastline as the California Current (Figure 1-2), all the heat that has been captured at the equator has been released to the air. What we experience is the cold water of the California Current that is flowing south back towards the equator to begin the cycle all over again. If you lived on the East Coast however, you would experience the warm water of the Gulf Stream that flows north near shore in the summer, bringing warm water up from the equator and the Gulf of Mexico.

The cold California Current is not the only factor in our cold water puzzle. Any beachcomber that has spent part of the summer in southern California and part in central or northern California will tell you that southern California water is definitely warmer. What's going on here? The answer is upwelling.

Upwelling is a process that occurs when the wind blows along the shoreline pushing the upper surface layer of water offshore and allowing the colder water from below to well up in its place (Figure 1-3). This process occurs all along the West Coast, however it occurs later in the season and much more strongly in central and northern California than in southern California. By late spring the winds that produce upwelling have pretty much abated in southern Cali-

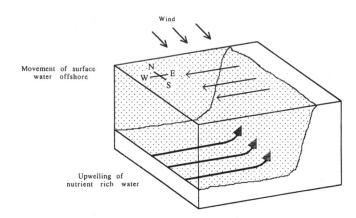

Figure 1-3. The upwelling process. Northwest winds blowing along the California coast move surface water offshore and cold, deep water replaces it.

fornia, allowing the nearshore surface water to stay in place and be warmed by the increasing spring and summer sunshine. In central and northern California however, the northwesterly winds that produce upwelling are just beginning in the spring, and will persist well into summer and sometimes into fall. Thus cold water nearshore is typical. And there is a bonus! As the air comes in contact with the ocean surface it becomes saturated with water vapor. The air is cooled when it comes in contact with the cold upwelled water, the water vapor condenses and produces that summer staple of central and northern California coastal existence . . . fog! As hot air rises from California's summer-warmed interior valleys, it pulls the fog onshore and sometimes it remains for weeks at a time.

Before you feel too sorry for us fog-bound beachcombers, a few other facts need to be made apparent. Upwelling allows water loaded with plant nutrients to enter the upper sunlit portion of the ocean, and when it does it fosters a tremendous level of photosynthesis in the single-celled marine plants called phytoplankton. This high level of plant production that persists through the spring and into the summer enriches the food chain greatly and produces one of the richest marine environments in the world. To the beachcomber this means a lot more diversity to discover and simply a greater abundance of marine life in general. Also the spring and summer fogs influenced primarily by upwelling tend to occur most persistently in the early mornings. During these months, the lowest low tides also occur in the early mornings. Therefore during the period when the fragile, low intertidal organisms of rocky habitats are liable to their most severe exposure to air, they are typically swathed in a cool blanket of moist, sun-blocking fog. In southern California, where the low tides occur on essentially the same early morning schedule, the fog is much less a factor and the intertidal zone is consequently exposed to the drying effects of warm air and sunlight. Does this take its toll of the organisms living there? Yes, it does, especially in exposed rocky intertidal habitats as we'll see.

THE CALIFORNIA MARINE FLORA AND FAUNA

Are Southern and Northern California That Different?

I grew up in southern California and have lived in the San Francisco Bay area since 1973. My answer to this question is a resounding YES, but with qualifications. As discussed earlier, there is a distinct difference in the degree of exposure to wave action north and south of Point Conception. Likewise, the differences in the length and positive effects of the upwelling season vary above and below this same geographic point. These differences in the physical nature of the marine environment certainly affect the composition of the community of organisms found along the length of the California coast. But, are these differences sufficient to really talk about distinct northern and southern California marine provinces? I don't think so.

The primary physical feature that influences the geographic range of a marine organism appears to be water temperature. Water temperatures vary north and south of Point Conception (Figure 1-2), and this geographic location is recognized as a distinct faunal boundary for many marine animal groups including fishes, snails, and clams. However, this barrier isn't 100% effective for any of these groups, and many of the intertidal zone organisms that a beachcomber will encounter are distributed continuously from Alaska to Baja California.

Origin of California's Marine Organisms

The scientific study of the distribution of organisms is known as biogeography. Marine biogeogra-

phers look at the geographic distributions of all the members of a group, and based on the pattern of overlap and co-occurrence they designate biogeographic provinces. Marine plants, especially the large seaweeds, essentially occur all along the California coast without distinct, recognizable provinces. For marine animals the biogeographers recognize a distinct cold-water province that includes much of Alaska, and a warm-water province that includes the Gulf of California and the tip of Baja California. In between these two is a province referred to as a cold temperate region, and most of California falls into this province. Over geological time water temperatures have fluctuated along our coast. During periods of warmer water, elements of the tropical fauna to the south have migrated northward. Conversely, when the water cooled down, elements of the Arctic fauna moved southward. When the sea water temperature began to change, these animals either adapted to the new water temperature or retreated back to their respective provinces.

What does all this mean to a beachcomber? Basically, it means when you explore a southern California habitat you will encounter some animals most of whose close relatives are in the warm-water province to the south. These animals usually also have a distinct northern limit to their distribution along the California coast. Examples of this would be the California spiny lobster (Figure 7-3) or the state marine fish, the Garibaldi (Color Plate 16b). Neither of these animals is common north of Point Conception and both have a majority of their closest relatives living in warm water. However, along with these restricted warm-water species, you will find a much larger group of organisms that is distributed along the entire California coast. Exploring a similar habitat in northern California will reveal a certain portion of the organisms with affinities to cold-water relatives and distinct southern limits to their distribution. These animals will occur along with essentially the same batch of hardy critters you saw in southern California.

To get back to the original question, "Are southern and northern California that different?" I would have to say yes and no. Southern California has a milder climate, seasonally warmer water, and more protection from direct wave action, along with more exposure to drying out during the low tides of spring and summer. Northern California is more exposed to heavy wave action and experiences a prolonged upwelling season along with the biotic richness, cold water, and fog that go along with it. However, the types of marine habitats that the beachcomber will encounter along the California coast are frequently much more influenced by the prevailing, unique conditions of the specific habitat rather than some overriding geographic condition. Likewise, the fact that most of the flora and fauna of these habitats is broadly distributed along the entire coast allows me to organize the beachcomber's guide into distinct habitat types that reoccur all along the Califfornia coast.

In this presentation I will introduce those organisms with the broadest distributions as well as point out species that are abundant only in the north or south of the state. It should be stressed that the organization of this book into distinct marine habitats is necessarily arbitrary. Most marine habitats are bordered by or merge with others and likewise the organisms found therein frequently overlap. Always be prepared for a surprise and don't hesitate to flip through the text to discover the identity of some creature that has the audacity to occur "out of place."

2
Keys to Successful Beachcombing

🐚

HOW AND WHERE TO LOOK

First of all I should point out that if you have taken up beachcombing, you're already a success! There is nothing more relaxing than strolling along a broad sandy beach in any kind of weather, or more exciting then the discovery of a new organism in an old familiar habitat. Here I will propose some helpful hints and suggestions for the beginner.

One of the first things new beachcombers should learn to do is adjust the way they observe the environment. We are preconditioned to look for large organisms by our own bulk and forget that most marine organisms are inches or less in size. Many of the most fascinating organisms will only be discovered when you are anticipating something small. The big things will take care of themselves.

Because many organisms are so small it is helpful to carry some type of magnifier. Many experienced beachcombers wear small hand lenses on cords around their necks. Others prefer a simple magnifying glass that they can slip into their pocket. Most nature-oriented stores stock a variety of such devices. One of the most ingenious is a plastic magnifier that is the size of a credit card and fits right inside your wallet.

Many of the small animals that become stranded on sandy beaches and some of the more delicate animals and plants of tidepools are more clearly viewed and appreciated when they are under water. Short of sticking your head in a tidepool, carrying a few small, sealable plastic bags can be very rewarding. A word of caution however, placing an animal in such a small volume of water and passing it around to be handled by warm human hands can quickly heat the water. Keep the observation time to a minimum or change the water frequently, every few minutes or so. Be sure to replace the animal where you found it. Don't carry glass containers, especially to rocky sites. Cuts from broken glass can bring a beachcombing outing to a quick and disappointing end.

Where To Look

Virtually any coastal site that is accessible along any of California's 15 coastal counties (Figure 2-1) can be a rewarding place to beachcomb. Be careful when descending coastal bluffs because many are unstable and a serious fall can result. Two of the most useful books available to the California beachcomber are the *California Coastal Access Guide* [1], and its companion text, the *California Coastal Resource Guide* [2], both prepared by the California Coastal Commission. These books provide maps and information on coastal access for all the coastal counties including details on parking, camping facilities and special points of interest. They are widely available at local libraries, as well as at nature and sporting goods stores, museums, and state parks.

As mentioned in the first chapter, rocky intertidal sites tend to have a greater diversity of organisms than do habitats with soft substrates. Sandy beaches tend to be the least diverse, but their exposed nature allows them to collect a continuing parade of interesting wash-ups and occasional visitors. This is especially true after a storm when the bounty of the ocean is sometimes augmented by the folly of human beings. A whole container of sports shoes was lost overboard off the Oregon coast in 1991, and the enterprising beachcombers soon had a cottage industry going that specialized in matching up pairs of washed-up shoes!

When beachcombing in soft sediment habitats, most discoveries will be made by digging. Sandy beaches virtually have no organisms that normally live exposed on the surface. Protected sand flats and mud flats may have a few hearty species obvious on the surface, but most organisms will be burrowed beneath the substrate. Carrying a small shovel along or even a garden spade that you can stick in your belt or day-pack can be very rewarding. (Remember to wash these implements, sea water is very corrosive!) Most organisms will be in the upper six inches of the substrate as they must maintain contact with the water to breathe. Some of the larger clam species in protected embayments may be considerably deeper, and only the most dedicated clammer is going to see them on any given day. Remember the size rule as you dig,

Figure 2-1. The fifteen coastal counties of California.

many of the most interesting organisms will be quite small. Dig a narrow trench about a foot deep, two feet long and six inches wide, and then carefully excavate the sediment along the side of the trench. Done with patience, this will reveal the organisms in their natural, burrowed position. When you are finished cover up your excavation and bury your discoveries under a few inches of sediment, they'll take it from there.

Beachcombing in rocky habitats can be a life-long adventure with new discoveries almost guaranteed with every trip. It is human nature to be drawn to the large and colorful organisms first. By all means enjoy them to your heart's content. However, once you've run out of big, obvious organisms, readjust your search pattern and explore the diversity of the smaller, more cryptic organisms. Many small animals seek the shelter of crevices or seaweed, and many live under loose rocks and boulders. Careful searching of any of these niches will be rewarded with new treasures. A word of caution about turning rocks. Animals live attached to the rock's bottom as well as freely underneath it. Be careful when turning the rock not to crush any of the inhabitants. When you've finished exploring, turn the rock back over the way you found it. Place the free-living animals under other rocks or among seaweed so they won't dry out. Remember organisms in the rocky intertidal zone are usually found in fairly specific locations relative to the tide and exposure. Don't move them out of their preferred locations.

Tidepools are also great places for beginning beachcombers to explore. Take your time. Find a pool a few feet across and about a foot deep. Sit down quietly beside it, hold still and watch. Soon you'll begin to detect flashes of movement from small tidepool fishes and shrimps. Next the hermit crabs will poke themselves out of their shells and resume their never-ending search for pieces of food. After several minutes the entire pool will come alive with the movement of dozens of snails and small crustaceans.

When To Look

Any time is a great time for beachcombing. Some of the most exciting trips occur at night when many nocturnal creatures are about. However, the serious beachcomber soon realizes that more can be seen during low tide than high, and some places aren't even accessible during high tide. It becomes necessary to learn about tides and tide tables to guarantee a fruitful beachcombing stint.

Tides are caused by the gravitational attraction of the sun and moon for the earth. The moon is the more powerful force as it is so much closer to the earth than the sun. The California coast experiences mixed, semi-diurnal tides. There are two daily high tides (one higher than the other) and two daily low tides (one lower than the other) over a 24-hour and 50-minute lunar day. Besides the daily cycle of tides, there is also a 29.5-day cycle that corresponds to the lunar month, the time it takes the moon to make a complete rotation around the earth. The sun and moon are lined up with one another during the new moon and full moon phases, and their combined attraction for the earth produces the most extreme tides, called spring tides. During the first and third quarter moon phases the sun and moon are at right angles to the earth, and their counteracting attractions produce the smallest tidal variations known as neap tides.

Tidal height on the West Coast is measured from an arbitrary zero point called mean lower low water (MLLW). This zero point is the average of all the lower low tides that occur in a year. All tidal levels on the West Coast are given in reference to this arbitrary point. A high tide listed in the tide table as +6.0 feet means the tide at its peak will be six feet above the average level of lower low water. A low tide listed as -1.5 feet means the tide will be a foot and a half below this average level. The latter tide is referred to as a "minus tide," and represents the best time to beachcomb as more of the intertidal zone will be exposed.

Tides are listed in most coastal newspapers and in yearly tables, which are available in most tackle shops or where fishing licenses are sold. The tides are given in reference to fixed geographic locations where tidal information has been gathered. Most tide tables provide corrections for points that fall in between the fixed tidal locations. For example, tide levels and times for the San Francisco Bay area are given for the Golden Gate Bridge. Consulting the tide correction table reveals that the high tide in Santa Cruz, which is down the coast, will occur one hour and nineteen minutes earlier than at the Golden Gate. At Coyote Point inside San Francisco Bay, the same high tide will occur one hour and fourteen minutes later than it does at the Golden Gate.

Tide tables can be made up years in advance because the position and effect of the sun and the moon are highly predictable. However, tide tables can not anticipate local weather conditions, which can significantly alter the actual tide that occurs on a given day. Large storm waves pushed by strong onshore winds can completely wash out a scheduled low tide, and cause a high tide to be several feet above the tide table prediction. Likewise, a strong high pressure area over the coast can push down on the water causing the low tides to be lower than predicted and the high tides to fail to reach their predicted heights.

The tidal cycle is just one aspect of local ocean conditions of which beachcomber should be aware. Periods of low wave action coupled with very low tides can provide the most opportune times for beachcombing, while high waves can prove very dangerous no matter what the tidal level. Never turn your back to the sea. Be very careful that you know whether the tide level is rising or falling. This is especially critical if you are exploring a cove or pocket beach that might get cut off as the tide rises. Similarly, a rocky intertidal area that has low-lying areas between you and the shore might flood as the tide comes in. Be very careful of large floating objects that lodge on the beach. This is especially true in northern California where large logs often accumulate on open sandy beaches. Never get between a log and the water on a rising tide. Common sense coupled with a knowledge of the local tides and an awareness of local ocean conditions can provide a safe, enjoyable day of beachcombing.

Where to Go When It's Raining

On those rare occasions when there is a six-foot-high tide, and it is blowing a gale, California offers the beachcomber continued access to marine life. To the south the San Diego Natural Museum in Balboa Park is a treasure trove of marine lore. Just north of San Diego, at the Scripps Institute of Oceanography in La Jolla, you can visit the brand new marine aquarium on the campus grounds. Los Angeles offers the Los Angeles County Museum of Natural History downtown, and the excellent Cabrillo Marine Museum and Aquarium in San Pedro. Central and northern California are twice blessed with the Mon-

terey Bay Aquarium in Pacific Grove, and the venerable Steinhart Aquarium at the California Academy of Sciences in San Francisco's Golden Gate Park.

A Note On Tidal Zonation

Beachcombing at low tide is usually best approximately two hours before and two hours after the scheduled low tide. Obviously, the lower the tide, the more area that will be exposed. This is important for rocky habitats as the organisms are distributed in somewhat distinct tidal zones. The lower intertidal zones are the more diverse as they experience less exposure to drying. In this guide I will present the organisms of the rocky intertidal habitat as they occur in these zones. A good beachcombing method is to explore the lowest exposed tidal zone first and move up into the intertidal as the tide rises. Remember to watch the waves!

GEAR FOR THE COMPLETE BEACHCOMBER

I have already mentioned several pieces of equipment that come in handy for beachcombers. Shovels, hand magnifying lenses, and plastic bags are all simple, but very helpful implements. I find it helpful to carry a small day-pack to stash my gear so my hands can remain free to explore. If you are beachcombing on a sandy beach in San Diego on a warm August afternoon, you may wish to wear nothing more than your bathing suit and sun block. More power to you! However, on that same afternoon in Crescent City or Half Moon Bay a down parka and hip waders might be more appropriate for the fog and cold wind that may occur.

A good dress code for beachcombing is that it is always better to over dress. You can always take it off if it gets too warm, but if you don't have it along, you can get very cold and ruin your day. Living in central California, my standard outfit consists of a faithful pair of jeans, a wool sweater, a wool stocking cap, and a 20 year-old army field jacket that I can't part with. I wear boot socks and a good pair of rubber hip waders. I carry a backpack with my camera, a hand lens, some plastic bags, sun block, waterproof paper and pencils and a baseball cap to exchange for the wool cap if it warms up. If it looks like rain, I substitute a yellow foul-weather parka for my field jacket or carry a large garbage bag as an emergency poncho. I seldom get too cold, and I can be down to a tee shirt in very short order if the weather calls for it. I also carry a change of clothes in my car for the periodic dunkings I invariably experience.

I'm not trying to create beachcomber clones here, just make the point that the day goes much more successfully if you are dressed for the weather. One final point is footwear. The classic barefoot beachcomber may be the ideal for the tropics and you may wish to emulate this on sandy beaches. However, you are better off shod when you explore rocky habitats that can have very sharp, slippery rocks. You will get your feet wet. It is required. Knowing this, you should avoid wearing leather shoes. Sea water will ruin even the best oil-treated boots. Wear shoes made of rubber or synthetic materials and rinse them well when you arrive back home. Soles that grip the surface, like the rippled soles on boat shoes, can give you some traction on slippery rocks.

CALIFORNIA'S FISH AND GAME REGULATIONS

Some beachcombers like to combine their exploration of California's coast with some fishing or clam digging. California offers many opportunities for these enterprises, but it is up to the individual to be aware of the sportfishing regulations that apply. In California anyone over the age of 16 years needs a license to take "any kind of fish, mollusk, invertebrate, amphibian or crustacean anywhere in the state." The only exception to this is that fishing from a public fishing pier in ocean or bay waters does not require a license. Fishing licenses are widely available in the state from bait shops, sporting goods stores, and even from many coastal drug stores. California has both annual and one-day licenses.

Regulations vary for the different species of sport fishes. Some are protected by size limits and closed seasons, and some have specific gear requirements. It is a good idea to check with the local tackle shop for specific regulations, or call the California Department of Fish and Game with your questions.

The rules governing the taking of shellfish are also quite specific. For certain species of clams you are required to take the first ten you encounter, while other species have a size limit. Again, it is a good idea to check the sportfishing regulations for the specific information. In addition to individual fishing regulations, clams, mussels, and scallops along the open California coast are quarantined from May 1 to October 1. This quarantine period corresponds to the time of the year when particular plankton undergo growth spurts and cause a condition known as a red tide. The red tide organisms (single celled plants called dinoflagellates) produce a toxin that is concentrated in the bodies of the filter-feeding mollusks like clams and mussels. Eating contaminated shellfish can result in a disease called paralytic shellfish poisoning in humans. Severe cases can be fatal. So be especially alert for posted notices of quarantines, and if there are any questions call the local health district office or the Department of Fish and Game.

A FINAL NOTE

I wish to conclude this chapter with a plea for conservation of our precious coastal resources. Every year the California coast is visited by millions of people seeking recreation, relaxation, and revitalization. Our very presence takes its toll. Mindful of this, each visitor should endeavor to leave the coast as intact as he or she found it for the next visitor. Be careful where you walk. Replace marine organisms carefully where you found them. Don't expose delicate marine animals to undue stress. The temptation to take a sea star or hermit crab home ends with the death of the organism and a smelly reminder of your folly. A photograph will last a lot longer and will be much more satisfying. Treat these coastal habitats with the same respect you would the home of a friend, because they are home to lots of friends of yours and mine.

References

1. California Coastal Commission. *California Coastal Access Guide*. 4th ed. Berkeley: University of California Press, 1991, 304 pp.
2. California Coastal Commission. *California Coastal Resource Guide*. Berkeley: University of California Press, 1987, 384 pp.

3
Some Quick Marine Biology*

NOTE ON SCIENTIFIC NAMES

Biologists by tradition give all unique living organisms a name. This is the name used to identify a distinct species, and it is referred to as the organism's scientific name. The scientific name consists of two words. The first, which is capitalized, is known as the generic name or the name of the genus to which the organism belongs. The second word, which is written in lower case, is the species or specific name. By convention, the scientific name is either written in italic characters or underlined. As you know, humans are classified as *Homo sapiens*. The common Pacific sea star of the rocky intertidal zone is *Pisaster ochraceus*. It is possible for two species to have the same generic name or to have the same species name (there are a whole batch of species named *californica*, for example), however, no two species will have the same genus and species names. Sometimes, there is some uncertainty to the species name of an organism, and it will be represented with its generic name and the abbreviation "sp." (e.g., a new *Pisaster* sp. was identified). Sometimes more than one species of a genus will be common to an environment, and they will be referred to as a group using the abbreviation "spp." meaning more than one species (e.g., several spp. were common).

In this book I will give the organism's common name or names first and will follow with the scientific name. Common names are easier to remember, and often much more descriptive, however they tend to vary with locality, and this can lead to confusion.

LIFE STYLES OF MARINE ORGANISMS

Unlike our earth-bound, terrestrial habitat the marine environment offers a new dimension for life to inhabit — the water column. True, we do have birds that fly in the air, but the water itself provides buoyancy and effortless transport that simply doesn't have a terrestrial parallel. Marine organisms have evolved into ingenious forms to take advantage of this resource. The organisms most beachcombers will encounter are associated with the bottom and remain in place when the tide goes out. We refer to these bottom dwelling organisms as *benthic* organisms or simply the benthos (meaning the organisms of the bottom). Animals and plants that use the water column for a home are referred to as *pelagic* organisms. Those pelagic animals and plants that drift passively with the water currents are known as *plankton*. Animal plankton is known as *zooplankton*, and the single-celled plants (diatoms and dinoflagellates) are called *phytoplankton*. The larger animals that are strong enough to swim free of the currents are called *nekton*. Nektonic animals include most fishes, squid and marine mammals.

Most organisms described in this book are members of the benthos. These are organisms that live on or in the bottom and remain behind when the tide recedes. However, most are very dependent on the water column for one or more of their vital requirements. Obviously, the organisms need the water to prevent death from dehydration, but many also use the water column for fertilization of their eggs, for transport of their kind to new environments, and for feeding. Sea urchins and sea stars, many snail species, clams, and a host of other benthic animals release their sperm and eggs into the water where fertilization occurs. The fertilized eggs typically develop into larval forms that reside in the plankton, developing and feeding, until they mature and settle back to the bot-

Quick and painless I hope!

tom. More sophisticated animals that have internal fertilization, like crabs and many bottom-dwelling fishes, still release their young as planktonic larvae.

Besides providing a rich food source for the planktonic larvae of benthic dwellers, the plankton and other organic material suspended and transported by water movement are a food source for a diverse group of benthic animals known as filter feeders. Animals such as sponges, clams, barnacles, sea cucumbers, and sea squirts employ an array of intricate filters to trap this rich food source. These animals are found in areas where water movement is relatively continuous, like the open sandy beaches and exposed rocky habitats.

When water enters a quiet protected area, like a closed embayment, factors that cause water movement, like wind and wave action, are reduced. As the water slows down, much of the suspended organic material falls out of the water onto the substrate. Here it is gleaned by a guild of animals known as deposit feeders, which include hermit crabs, some species of snails and shrimp, and many different kinds of worms. Some deposit feeders pick up individual deposited particles with specialized appendages and mouth parts. Others simply engulf the substrate wholesale like earthworms, and digest whatever organic matter it contains. Deposit feeders tend to dominate areas of quiet water in bays and estuaries where the bottom is typically soft mud.

Intertidal and nearshore subtidal rocky areas provide the solid substrate necessary for the attachment of large marine plants. These are chiefly the green, red, and brown seaweeds and surfgrass. Seaweeds provide a food source of a large group of invertebrate herbivores, such as snails and chitons, that graze directly on the living plants. Pieces of seaweed that are detached by wave action or herbivores provide a source of drifting algae that forms the staple diet of animals like sea urchins and abalone. Finally the large plants are broken down mechanically by water movement and by decomposers like bacteria into particulate matter that becomes food for filter and deposit feeders.

With all the filter feeders, deposit feeders and herbivores around, it's not surprising that an array of carnivorous species are in place to eat them. Fish and squid move in from the water column, while sea stars, crabs, and some snails and worms search along the various bottom habitats for their prey.

MAJOR MARINE PLANT AND ANIMAL GROUPS

Beyond the scientific name, biologists have erected a hierarchy of scientific classification. Don't panic. I will make this simple and list the major marine plant and invertebrate animal groups with a brief description of each. For plants this includes the major marine seaweed groups and the few marine grasses that occur in California. Marine invertebrate animals are incredibly variable in size and form. Here, they are introduced at the classification level known as the phylum (plural phyla), which unites broadly related invertebrates into a general group. The very diverse marine invertebrate phylum, Mollusca, is further described at the more concise classification level known as the class.

The vertebrate animals in the sea include the fishes and marine mammals. When accessible to beachcombers, most marine habitats will include few obvious fishes, so only a handful are included in this guide. There are several field guides to marine fishes available [1-6]. Marine mammals tend to be a wary lot and will usually only be viewed from some distance. The more common species seen in California are treated in a Chapter 8 at the end of this book. In addition there are several good marine mammal references available [7, 8].

Major Marine Invertebrate Phyla

Phylum Porifera — the sponges. The sponges are the simplest of the multi-celled marine invertebrates. Their bodies are basically porous filters that trap organic material down to the size of bacteria. The sponge has many small openings or pores over its entire surface that lead to chambers where filtration occurs. These chambers are lined with special cells, called collar cells, that possess long, whip-like flagella. These flagella beat to create the current that pulls water into the sponge and sends it out again through larger exit openings. Most sponges in the intertidal zone grow close to the rocky surface and spread in an encrusting sheet of sponge tissue. In deeper, quieter water sponges grow erect and reach elaborate sizes and shapes in the more favorable habitats such as coral reefs. Sponges are often brightly colored with red, orange, and yellow species being common.

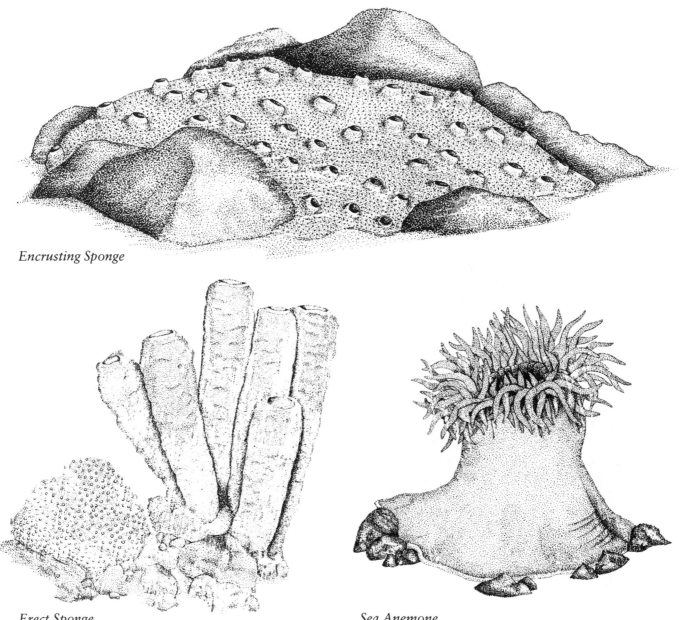

Encrusting Sponge

Erect Sponge

Sea Anemone

Phylum Cnidaria. The cnidarians include such familiar animals as sea anemones and jellyfish. (By the way, the "c" is silent, so it's pronounced "nidarians.") All members of this phylum possess special cell organelles called nematocysts, which are used to capture food and for defense. Nematocysts come in a variety of types and some penetrate and poison the prey while others entangle it with sticky substances. The nematocyst-bearing cells are always located on a ring of tentacles surrounding the cnidarian's mouth, and elsewhere on the body depending on the group.

There are two basic body types found in the cnidarians, the polyp and medusa. The sea anemone is an example of the polyp body form. It consists of a column of tissue with a bottom disk for attachment and an upper disk that bears the mouth and tentacles. Polyps are attached to the substrate and often occur in colonies of many polyps such as a coral colony. The polyps in a colony can be quite small in size, and the over-all colony bush-like. These bush-like colonies are known as hydrozoan colonies or hydroids.

The jellyfish is an example of the medusa body form. It has a bell-shaped umbrella with the mouth hanging downwards in the center, surrounded by a ring of tentacles around the edge of the umbrella. The medusa can swim by contraction of the bell, which

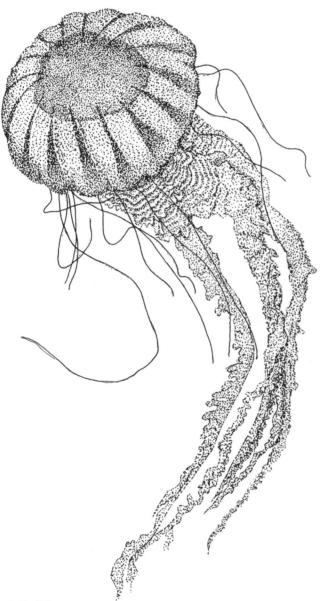

Jellyfish

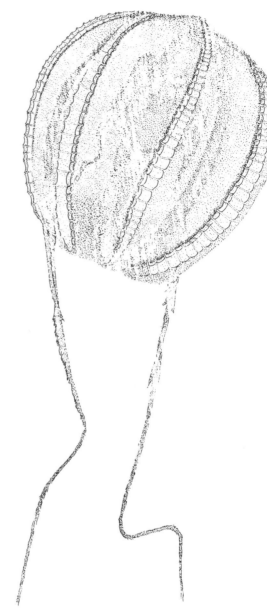

Ctenophore

forces water out and propels the animal in the opposite direction. Some cnidarians have both body forms in their life cycles while others have only the polyp or the medusa form.

Phylum Ctenophora. Ctenophores are a small group of medusa-like animals that are commonly called comb jellies. They are considered to be closely related to the cnidarians as they have a circular body plan, but they do not possess nematocysts. Ctenophores are planktonic animals. They swim through the water by the coordinated beating of eight bands of fused ciliary plates called ctenes. Ctenophores feed on planktonic animals, which they

either capture with special adhesive tentacles or swallow whole. They are sometimes quite abundant in bays and often wash up on sandy beaches.

Phylum Platyhelminthes. The name literally means "flatworm," and includes the major parasitic groups, the flukes and tapeworms. It also includes a large group of simple, free-living worms that vary in size from microscopic to two feet in length. The marine flatworms typically encountered by the beachcomber are found under smooth rocks in the intertidal zone. These animals are usually a mottled brown or gray color, from one to two inches long, and are very flat. They are about three times as long as wide and

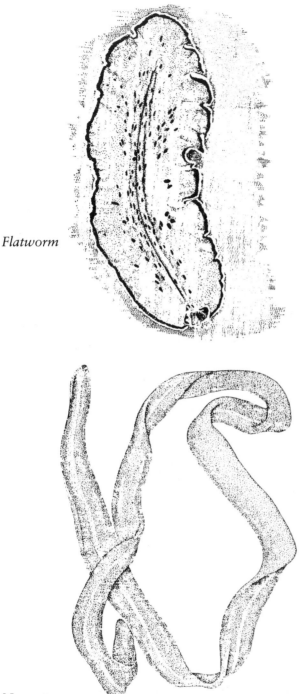

Flatworm

are considered to be closely related to the flatworms. However, instead of being flat they are very elongated. The long worm body is adapted to moving through tight quarters, burrowing through the sediment or living in a tube. Ribbon worms vary in size from a few inches to several feet in length, and are capable of tremendous stretching. A worm eight inches long can stretch to over a yard long! Ribbon worms are found in a variety of habitats including soft sediments and the rocky intertidal zone. Most make their living feeding on live invertebrate prey, which they capture with an unique feeding structure known as a proboscis. The proboscis is housed in a special cavity in the body wall and can be shot out very rapidly by the contraction of the body wall musculature. The proboscis is sticky or armed with poison barbs, and quickly entangles and subdues the prey, which is then brought to the mouth and swallowed.

Phylum Annelida. The annelid worms include the familiar garden earthworm and the medicinal leech.

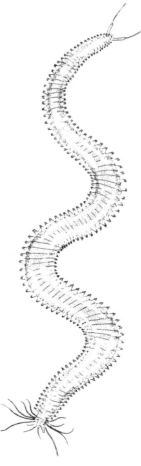

Most marine annelids belong to the annelid class Polychaeta. Polychaete worms are widely distributed in the marine environment and have a variety of life styles. Several species will be discussed in this guide. Shown here is a free-living or "errant" polychaete from the genus *Nereis*. As you can see *Nereis*' body is organized in distinct rings (annelid means arranged in rings) or segments, each equipped with special appendages for locomotion. These segmental appendages are strengthened by stout bristles or setae, which assist the worm in gaining traction with the substrate. The head is equipped with several sensory appendages for testing the worm's envi-

Nemertean

often have a series of eyespots anteriorly and along their sides. These flatworms glide along the moist rocks on cilia and can also swim by undulating their flat bodies. Intertidal flatworms have a mouth located in the center of their bottom (ventral) surface and feed on organic debris or small invertebrates.

Phylum Nemertea. The nemertean worms are known as ribbon worms or rubber-band worms. They

Errant polychaete Nereis

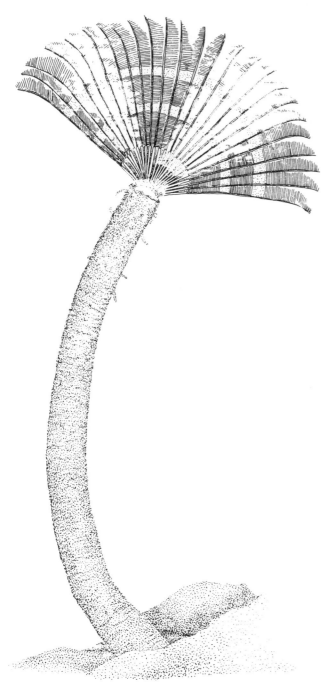

Sedentary polychaete, featherduster worm

featherduster worm. This polychaete lives in mucous tube and the elaborate plume of "feathers" are special ciliated head appendages, which are used for filter feeding. Sedentary polychaetes typically live in a permanent or semi-permanent space like a tube or burrow. They feed on particulate organic matter or plankton, which they capture by filter or deposit feeding.

Phylum Sipuncula. The sipunculid worms are commonly called peanut worms. They are considered to be close relatives of the annelids. Sipunculids are simply organized with a body consisting of a bulbous trunk region and an elongated "neck" or introvert, which they can roll inward like the finger on a glove. When the introvert is rolled in all the way, the swollen trunk has the shape and size of a peanut, and

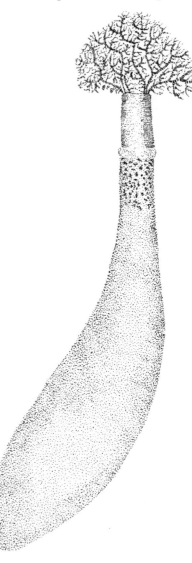

Sipunculid/peanut worm

ronment as it encounters it head first. *Nereis* is able to evert its pharynx (the anterior region of its digestive tract), which is armed with jaws to capture prey. Errant polychaetes typically move about freely either through soft substrates or on the surface, and feed on live prey or large food particles like seaweed or carrion (dead animal tissue).

The second polychaete illustrated here is an example of a "sedentary" polychaete, and is known as a thus the origin of their common name. When the introvert is unrolled (everted), the mouth opening is found at the end surrounded by a ring of tentacles, which are used for filter or deposit feeding. Sipunculid species are found in soft substrates among the roots and holdfasts of surfgrass and seaweeds, respectively, and also in cracks and crevices among the rocks in the rocky intertidal zone. The common species found along California's coast are small, ranging in size from one to four inches long when the introvert is fully everted.

Phylum Echiura. The echiuran worms are also considered to be closely related to the annelids. These worms are very plain looking with a bulbous trunk and an anterior, flexible proboscis that contains the mouth. Echiurans are found in soft sediment habitats and are usually deposit feeders. The only species that will be commonly encountered by California beachcombers is the fat innkeeper worm, *Urechis caupo*,

living in the sandy-mud flats of coastal embayments. The innkeeper, shown here, is a burrow-dwelling, filter feeder.

Phylum Bryozoa. The bryozoa are small, colonial animals that are often overlooked. These miniature animals live in small calcareous (made of calcium carbonate) houses that are attached together. The houses are usually rectangular in shape, and a colony consists of large numbers of these small houses to give an overall basket-weave or reticulate pattern.

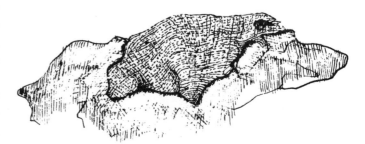

Bryozoan colony — flat, encrusting sheet

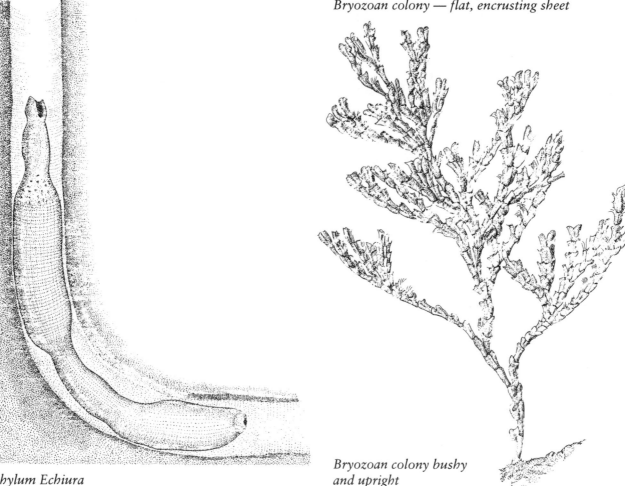

Phylum Echiura

*Bryozoan colony bushy
and upright*

Bryozoans are filter-feeders. They employ a tentacular device known as a lophophore for feeding, which can be elevated above the calcareous house. A variety of colonial growth forms is found among bryozoan species. Some colonies grow in flat, encrusting sheets on a variety of substrates including rocks, the shells and exoskeletons of animals, and on seaweeds. Other colonies grow bushy and upright and look superficially like finely-branched algae.

Phylum Phoronida. The phoronids are a small group of marine worms that also have the feeding structure known as a lophophore like the bryozoans previously discussed. The phoronid lophophore is a

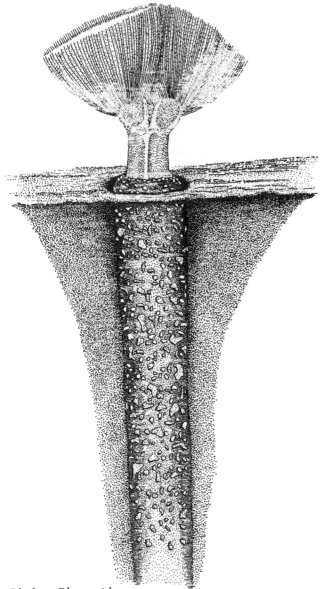

Phylum Phoronida

continuous double row of ciliated tentacles folded into an elaborate horseshoe shape. The phoronid has a plain, unsegmented body that remains enclosed in its chitinous (stiff, cellulose-like material) tube. The tube is oriented vertically in the substrate with the lophophore extended anteriorly above the surface for filter feeding. A common California species is *Phoronopsis viridis,* which lives in large aggregations in the low intertidal regions of sandy mud flats in protected coastal embayments.

Phylum Mollusca. The mollusks are one of the largest and most diverse of the invertebrate phyla. Included are such different forms as clams, squid and abalone. Mollusca means soft body in Latin, and at the heart of the molluscan organization is a soft, pliable body protected by a calcareous shell. The shell is secreted by a special epidermal tissue known as the mantle, which completely underlies the shell. Between the shell and the body is a space, lined by mantle tissue, that houses a pair of large gills used for breathing. This is called the mantle cavity. The basic mollusk hauls this shell-encased soft body around on a broad, flat muscular foot. The head is at the anterior end of the body and typically has a pair of sensory tentacles, and a special feeding structure known as the radula within the mouth cavity. This is a description of what zoologists visualize as an ancestor to the modern mollusks, and it would look similar to a modern-day marine snail we call a limpet. Three limpet shells are illustrated here. Snails belong to the class Gastropoda, which contains a wondrous number of species with and without shells, many of which will be found in this guide.

Not too far removed form the basic mollusk just described is the chiton (class Polyplacophora). The chiton has a shell divided into eight valves or plates. This gives it some flexibility and allows it to conform and adhere to the irregular surfaces of the intertidal rocks where it typically lives. Chitons still possess the large molluscan foot for locomotion and adhesion, and a large radula for feeding on seaweeds.

The class Bivalvia includes all the mollusks that have a shell divided into two plates or valves that completely encase the body, such as a clam or a scallop. All bivalve mollusks have a body that has been compressed laterally (from the sides) and a foot that is now spade-shaped and narrow instead of broad and flat. The shell is divided in two by a hinge made of protein along the mollusk's back, which allows it to gape open at the bottom so the foot can protrude. The spade-like foot is used for burrowing into the substrate for safety. Bivalve mollusks have lost the radula and instead use their large ciliated gills to filter food. As filter feeders, the bivalves can feed anywhere they can successfully attach or burrow and have access to the water column. Thus such diverse forms as clams,

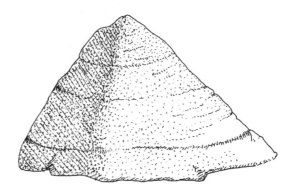

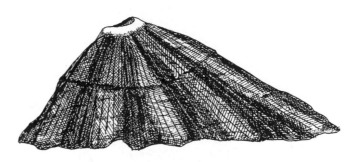

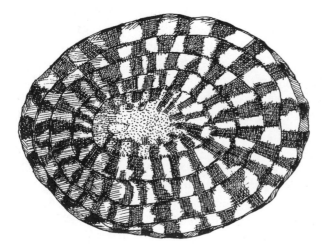

Snails called limpets

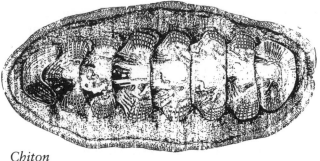

Chiton

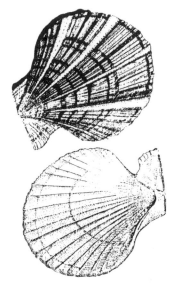

Scallops

Octopus

mussels, oysters, and the scallops shown here can be found.

Finally comes the class Cephalopoda, which includes the familiar octopus and squid. Cephalopods are the most intelligent and motile of the mollusks. Octopus move along the bottom on eight, suckered arms and swim via jet propulsion by forcing water in and out of their enlarged and muscled mantle cavities. Squids live in the water column, and their mantles are even larger and more muscular than that of the octopus, giving them a rapid swimming style that lets them swim and capture some fish. These animals are carnivores. They feed with stout, beak-like jaws that allow them to subdue prey and bite them into pieces.

Phylum Arthropoda. The largest of all the animal phyla is the Arthropoda. Included here are the insects, the spiders, and in the marine environment, the crustaceans (class Crustacea). The arthropods have two morphological features that are the keys to their incredible success. The first is a thickened outer cov-

ering called an exoskeleton. The second is the jointed limb (arthropod literally means jointed limb). The limb is actually a series of tubes joined together by flexible membranes. Each of these joints allows movement in only one plane. When a number of these tubes are strung together in line with muscles to flex them, and each joint flexing in a different plane, a limb with a great range of movement can be obtained. Thus we find arthropods that can walk, swim, and even fly.

The marine arthropods known as crustaceans occur in a bewildering number of shapes and sizes. Such diverse animals as barnacles and shrimp are both crustaceans. A review of the group would be too lengthy here, and the individual crustacean species will be introduced separately in the guide. However, it should be remembered that all these animals have the protective exoskeleton and the jointed limb in common. The exoskeleton is made of a tough animal material called chitin and calcium carbonate salts that give its "crusty" texture. Also all crustaceans have the necessary limitation the exoskeleton imparts in that they must shed the exoskeleton to grow. The periodic shedding of the old exoskeleton and its replacement with a new one is known as molting. Molting is very

Crustacean — barnacles

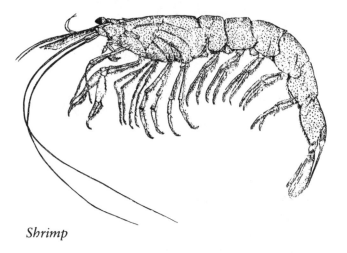

Shrimp

bers. Echinoderms have an internal skeleton like our own, composed of calcium carbonate. The individual skeletal elements may be tightly joined together like the skeleton or "test" of a sea urchin, or be more loosely articulated to allow movement like that seen in the arms of sea stars or serpent stars. The tube feet are equipped with suckers in the sea stars, sea urchins and sea cucumbers, and are used for locomotion as well as food gathering. The serpent stars use their long, thin arms to drag and push themselves along, and their tube feet serve mainly for food manipula-

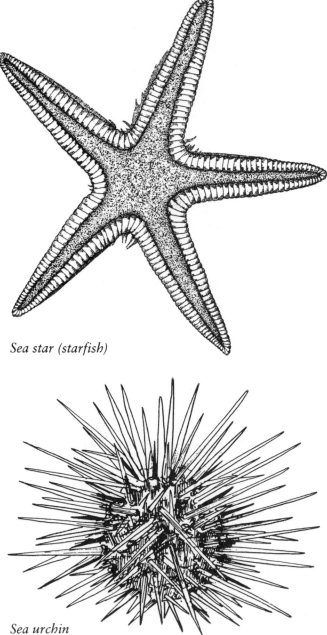

Sea star (starfish)

Sea urchin

time and energy consuming, and leaves the crustaceans very vulnerable to harm. The new exoskeleton must be soft so it can be extracted from the old one, and to allow it to be stretched so the animal can increase in size. The crustacean is defenseless while it is molting and waiting for the new exoskeleton to harden. During this time they become very secretive.

Phylum Echinodermata. The echinoderms are a group of exclusively marine animals. They are all organized in a circular or radial fashion as adults and possess a unique, internal hydraulic system known as the water vascular system. The water vascular system uses sea water as its hydraulic fluid and terminates in structures known as tube feet. Tube feet are used for locomotion, feeding, and respiration (breathing). Echinoderms include the familiar sea stars (starfish) and sea urchins. Also found in this phylum are the serpent or brittle stars and the elongated sea cucum-

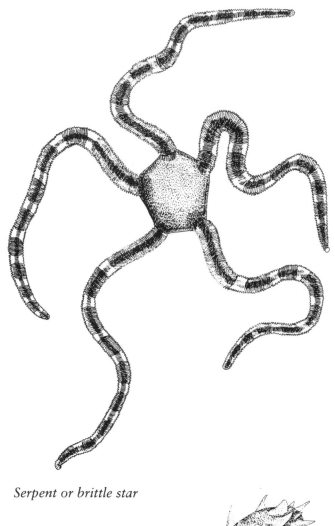

Serpent or brittle star

tion. All echinoderms use the thin-walled tube feet as sites for respiration. As a general rule, echinoderms are relatively large invertebrates and species will be treated individually in this guide.

Phylum Chordata/Subphylum Urochordata. The chordate animals have a dorsal hollow nerve cord, a stiff rod known as a notochord, gill slits, and a post anal tail. This phylum includes the familiar back-boned animals including ourselves (subphylum Vertebrata), and the subphyla Urochordata and Cephalochordata known as the invertebrate chordates. Only the Urochordata will be treated here.

Most animals in the subphylum Urochordata are sessile (stationary) filter-feeders known as sea squirts or tunicates. Sea squirts look very little like other chordate animals as adults and only their larvae show the chordate characteristics previously mentioned. Sea squirts vary in size from peach-sized individuals to minute animals joined together in a colony. The sea squirt is encased in a skin-like tunic, which can be thick and woody or soft, transparent, and quite colorful. Within the tunic the body of the sea squirt is dominated by a large filter-feeding basket, which is the forerunner of our own throat region. Water is pumped into and through openings or slits in the basket by

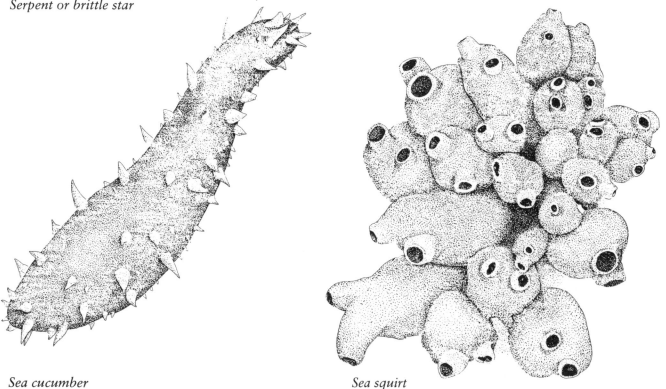

Sea cucumber

Sea squirt

cilia on its inner surface. Food particles are trapped as they pass through these openings, and are carried on to the animal's stomach. Sea squirts are usually attached to a solid substrate. Large solitary species are typically attached to rocky substrates. Colonies of miniaturized individuals living in a common tunic may be found on a variety of surfaces including rocks, pier pilings and floats, as well as the bodies of other invertebrate animals, such as the shell of a mussel as illustrated here. These colonies are known as compound tunicates.

Marine Plants

Most large, obvious plants a beachcomber will see are seaweeds or algae. The seaweeds are considered relatively primitive plants by botanists because they lack the elaborate conductive tissues and nutrient-gathering roots of the vascular plants that dominate terrestrial environments. However, because seaweeds are bathed by nutrient-bearing water, they can survive perfectly well without these specializations, and many grow to very large sizes.

Seaweeds require light and a solid attachment site to prosper. They are attached to the substrate by a mass of root-like, adhesive processes collectively called a holdfast. Because the entire plant is capable of absorbing the nutrient fertilizers from the water, true roots are not necessary, and the holdfast tissue has no special absorptive ability. In a typical alga, stem-like structures called stipes grow from the holdfast. The stipes in turn support flat blades. In some seaweeds there are no recognizable stipes, and the blades grow directly from the holdfast. The relative size of the stipes and blades vary considerably from one algal species to another. The size of the holdfast is directly related to the size of the plant and the particular environment in which it grows. Algae growing in the exposed rocky intertidal zone need much larger holdfasts than those growing in a protected, quiet water environment.

Marine seaweeds belong to three main groups, the green algae (Chlorophyta), the red algae (Rhodophyta), and the brown algae (Phaeophyta). Green algae are the least diverse and tend to be found in fairly shallow, inshore environments. The plants are usually thin, sheet-like and a bright green color. Green algae can be very abundant in a local situation and very short-lived. For example, the tissue-thin sea

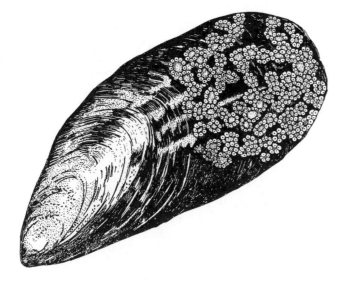

Colonies of miniaturized sea squirts living in a common tunic on a mussel shell

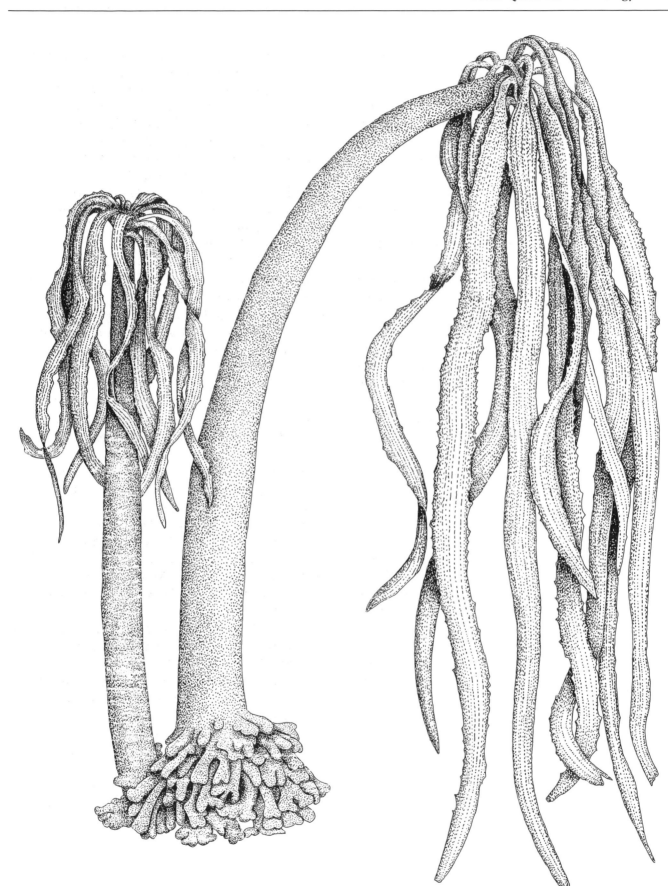

A typical alga — Postelsia palmaeformis

lettuce, *Ulva* spp. (Color Plate 1a), sometimes covers nearly every hard substrate visible, and a week or so later is almost entirely gone.

Red algae are the most diverse, and individual species occur in a range of colors and sizes (See Color Plate 2). A red alga can be red, brown, green, or violet. Species vary in size from intricately-branched, lacy forms a few inches long, to broad, flat blades over two feet in length. Some red algae, called coralline algae, incorporate large amounts of chalk-like calcium carbonate between their cell walls and have a coral like texture. Red algae occur from the intertidal zone down to over a 100 feet below the surface.

Brown algae are usually brown to dark brown in color. Among them are the largest of all the seaweeds, the kelps. Kelp species can reach well over 100 feet in length, and occur in large subtidal aggregations or beds. Kelp species occur from the middle rocky intertidal zone down to below 100 feet. Smaller, non-kelp brown algae are found in the high rocky intertidal zones.

The beachcomber might encounter several hundred common marine algae in the California intertidal zone. Only the largest, most obvious species are mentioned in this guide. There are several more extensive guides available [9, 10].

The so-called higher plants or vascular plants are not widely represented in California marine environments. Surfgrass (Color Plate 3b) is found in the middle and low rocky intertidal zones and below, and eelgrass (Color Plate 3c) occurs in quiet water habitats. There are a number of common salt marsh plants, and plants that occur on coastal dunes. Beachcombers can find more extensive information about these plants in the following references [10, 11, 12].

This completes a brief introduction to marine biology. Beachcombers wishing to find more in-depth information on the subject are referred to the following texts [13-15].

References

1. Eschmeyer, W.N. et al. *A Field Guide to Pacific Coast Fishes of North America*. The Petersen Field Guide Series. Boston: Houghton Mifflin Company, 1983, 336 pp.

2. Fitch, J.E. and R.J. Lavenberg. *Tidepool and Nearshore Fishes of California*. Berkeley: University of California Press, 1975, 156 pp.

3. Fitch, J.E. and R.J. Lavenberg. *Marine Food and Game Fishes of California*. Berkeley: University of California Press, 1971, 179 pp.

4. Miller, D.J. and R.N. Lea. *Guide to the Coastal Marine Fishes of California*. Sacramento: California Department of Fish & Game, 1972, 249 pp.

5. Gotshall, D.W. *Fishwatchers' Guide to the Inshore Fishes of the Pacific Coast*. Monterey: Sea Challengers, 1977, 108 pp.

6. Ferguson, A. & G. Cailliet. *Sharks and Rays of the Pacific Coast*. Monterey: Monterey Bay Aquarium Foundation, 1990, 64 pp.

7. Leatherwood, S. and R.R. Reeves. *The Sierra Club Handbook of Whales and Dolphins*. San Francisco: Sierra Club Books, 1983, 302 pp.

8. Orr, R.T., & R.C. Helm. *Marine Mammals of California*. Berkeley: University of California Press, 1989, 92 pp.

9. Abbott, I.A. and G.J. Hollenberg. *Marine Algae of California*. Stanford: Stanford University Press, 1976, 827 pp.

10. Dawson, E.Y. and M.S. Foster. *Seashore Plants of California*. Berkeley: University of California Press, 1982, 226 pp. .

11. Barbour, M.G. and J. Major. *Terrestrial Vegetation of California*. New York: John Wiley, 1977, 1002 pp.

12. Munz, P.A. and D.D. Keck. *A California Flora and Supplement*. Berkeley: University of California Press, 1975, 1905 pp.

13. Niesen, T.M. *The Marine Biology Coloring Book*. Philadelphia: Barnes and Noble, 1982, 209 pp.

14. Nybakken, J.W. *Marine Biology, An Ecological Approach*. New York: Harper and Row, 1987, 514 pp.

15. Sumich, J.L. *An Introduction to the Biology of Marine Life*. Dubuque: Wm. C. Brown, 1992, 449 pp.

4
The Sandy Beach

TYPES OF SANDY BEACHES

Sandy beaches occur all along the length of the California coastline (Figure 4-1). They represent the most physically controlled of all the nearshore marine habitats, and as such one of the most difficult to live in. Some background on the formation and dynamics of sandy beaches is helpful in understanding the nature of the habitat and the animals that live there.

Sandy beaches have several things in common. All are composed of sediment particles that overlay a rocky beach platform. Beyond that they can vary considerably. The sediments that make up California's beaches are mainly small pieces of the minerals quartz and feldspar. These minerals are either weathered from the continental rocks and washed down to the ocean in rivers or are the products of coastal erosion. Once these sediment particles reach the sea, they are carried along the coast until they find their way onto a sandy beach or offshore and down a submarine canyon. The transport of sediment along the coast requires water movement in the form of waves and currents.

Waves come from all directions along the California coast, but the prevailing direction is from the northwest. When the waves come on shore and break, they often break at an angle to the shoreline with some of their energy generating currents that run along the beach. These are called longshore currents and as they move along the coast they carry sand particles with them. This is called longshore transport (Figure 4-2). Because the prevailing wave direction is from the northwest, the prevailing longshore current flows towards the south and the net longshore transport of sediment is likewise southerly.

The larger the wave, the more energy it contains and the bigger the sand particles it can carry. As

Figure 4-1. Waves wash up the face of a California sandy beach.

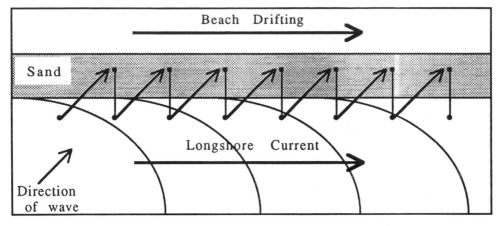

Figure 4-2. Direction of longshore transport of sediment along a beach with a prevailing northwest swell.

waves loose their energy, the larger, heavier particles can no longer be carried along and fall to the bottom. The same is true for any type of water movement. The faster the movement, the larger the particles that can be transported. As the water slows down, the larger particles can no longer be held in suspension and fall to the bottom. Finally, when the water becomes still, only the smallest, lightest particles remain in suspension and available to settle to the bottom. These small, light particles, known as silts and clays, are deposited as marine muds either in protected, quiet habitats like bays and estuaries or farther offshore in deep water. This settling process is known as sedimentation, and it accounts for the formation of all soft bottom marine habitats.

Because sandy beaches are subject to relatively continuous wave action, only the larger particles, known as sands, accumulate and remain on the beach. The size of the sand particles on the beach is directly related to the size of the waves that hit the beach. Beaches that experience vigorous wave action will have large sand grains and feel coarse to the touch. Beaches that are exposed to gentler wave action will have finer particles, such as the sugar-like sand of Carmel Beach on the Monterey Peninsula. Beaches will also undergo drastic changes in the type of sediment present. Strong storm waves can strip a beach of sand or leave only the largest particles behind. Likewise small waves return fine grains to the beach, creating a gentle, sloping beach. Many beaches go through an annual cycle of sand accumulation during the spring and summer months. This is followed by a drastic loss of sand with the first winter storm, which leaves the beach in a winter condition [1].

Sand washed down from rivers and carried along the coast by longshore transport accumulates on rocky benches called beach platforms (Figure 4-3). These beach platforms have been carved into the rocky continent by wave action. The size of the beach platform depends on the local geological make up of the coastline. Small, narrow wave-cut benches are found at the base of towering headlands made of erosion-resistant rocks. Long, broad beach platforms are cut from softer rocks. In some areas of the coastline it is possible to view a series of extensive beach platforms, also called beach terraces, that rise up like stepping stones above the current sea level. These represent wave-cut terraces that were carved during periods when the sea level was higher.

In some areas of California the sand that accumulates on the beach is blown shoreward off the beach face and forms dunes behind the beach. These dune fields can be extensive, such as those found in Humboldt and Monterey counties. The beautiful Golden Gate Park in San Francisco is built almost entirely on a sand dune field!

Sandy beaches are truly dynamic habitats. What's it like to live there? First there is the sediment to contend with. It doesn't stay put; therefore living on the surface or trying to excavate and live in a permanent burrow are not options. Because the sediment can shift so rapidly and unpredictably, any organism living on the beach must be able to swim and/or burrow rapidly to keep from being swept away. Then there is a problem with drainage. As the tide recedes, water drains from between the sand grains, leaving organisms burrowed in the sand in danger of drying out. This is less of a problem on beaches made up of fine sand grains as water is held in the small spaces between the grains by capillary action. The spaces between the grains on coarse grained beaches are too large to hold water by capillary action. These beaches

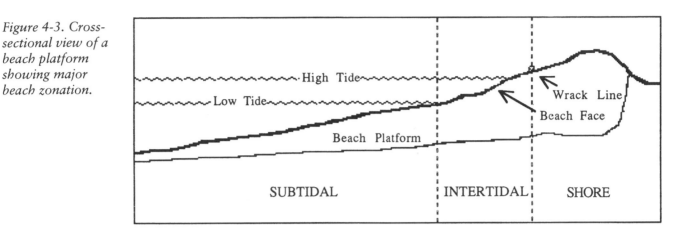

Figure 4-3. Cross-sectional view of a beach platform showing major beach zonation.

High Tide

Low Tide

Wrack Line

Beach Face

Beach Platform

SUBTIDAL | INTERTIDAL | SHORE

can become quite dry at high tide, especially if the sun is shining and/or the wind is blowing. Finally there are predators. Fishes, crabs, shrimps, worms, and an occasional predatory snail forage on the beaches at high tide. Birds take over at low tide.

If a sandy beach is such a miserable place, why live there at all? Because there is food there, plain and simple. The same wave and current action that accounts for the beach's dynamic physical nature also delivers food to the sandy beach. Some of the food consists of the carcasses of large animals like fish and marine mammals, or large seaweeds that have been torn from their attachment offshore. However, most of the food that is available to beach dwellers is in the form of small particles of animal and plant material called detritus. These particles are derived from large and small organisms produced elsewhere in the ocean that have been broken up and carried along by the moving water.

PERMANENT BEACH RESIDENTS

Crustaceans

Given the physical description of the sandy beach habitat and the nature of the food available, allow me to introduce the ultimate sandy beach animal, the mole or sand crab, *Emerita analoga* (Figure 4-4). The mole crab is found on sandy beaches from British Columbia to Baja California. The body is oval and streamlined for easy burrowing, and the hind limbs are modified for rapid digging into the sand. Mole crabs burrow backwards into the sand facing down the beach into the oncoming waves. However, they

don't stay put. As the tide ebbs and floods, the mole crab emerges from the sand and rides the tide up or down to position itself in the swash zone. The swash zone is that area of the beach washed by both the breaking waves rushing up the beach and the water flowing back down the sloping beach. Here in the swash zone the mole crab unfurls a pair of long antennae covered with fine hairs. The antennae are held up into the water and the hairs trap detrital particles. The antennae are then wiped across the mouth and the trapped particles are removed and swallowed. If a particularly strong wave should dislodge the mole crab, it is a strong swimmer and quickly finds the bottom and re-burrows.

Where to find mole crabs? By digging in the swash zone of course. The crabs will be just beneath the surface and can be easily dug up with your bare hands. However, if they are particularly abundant on the beach some individuals invariably become stranded by the receding low tide and can be found on the exposed beach in the moist sand several inches below the surface. Mole crabs are crustaceans and must shed their outer skeletons to grow. The heavy shield-like portion of the skeleton, called the carapace, is frequently found by beachcombers on the surface of the sand.

Mole crabs are a quarter of an inch long when they settle from the plankton in the spring and summer. Male crabs stop growing after they reach about three quarters of an inch and live only one year. The largest female crabs grow to over two inches and can live two years. Large females may be brooding embryos. Turn a female over and look for the bright, coral-colored mass of fertilized eggs on her underside. A female will brood her eggs for several weeks until they develop into planktonic larvae. Then the larvae

Figure 4-4. Top and profile views of the two-inch-long mole crab, Emerita analoga; *upper: crab with antennae in feeding position.*

break through the egg membrane and swim off into the sea. There they feed on plankton and grow until they are ready to settle and start life on the beach.

The shallow holes you excavate while digging in the swash zone for mole crabs often fill with water and reveal another sandy beach crustacean known as a mysid (Figure 4-5). You may get a glimpse of this small animal (less than one half inch long) swimming in the water and then settling to the bottom and quickly burrowing. Agitating the sand a bit will cause it to resume swimming. Mysids are also known as opossum shrimp because they carry their eggs in a pouch located on their trunk region. They are nearly transparent, and sometimes are more easily found by looking for their shadows on the bottom of the hole.

Mysid species occur in several marine habitats including tidepools and in estuaries.

Another common sandy beach crustacean is the isopod, *Excirolana* sp. (Figure 4-6), a close relative of *Cirolana harfordi* from the rocky intertidal zone. *Excirolana* is sometimes mistaken for a small mole crab because it has a similar, oval appearance. However, it has much shorter antennae and body appendages, and only reaches about one half inch in length. What it lacks in size it makes up for in number and appetite! *Excirolana* is a scavenger and hundreds to thousands of these animals can be found attacking any large animal carcass that might wash up on the beach. It will quickly turn a dead fish into a skeleton, and some waders claim they've been nipped by the

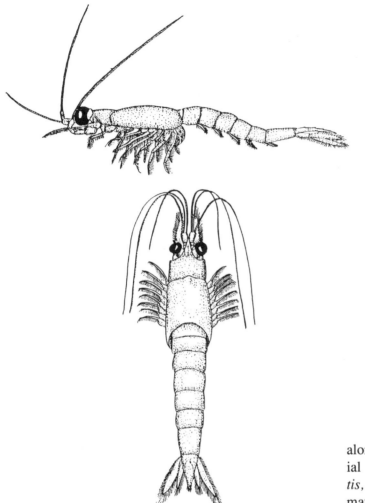

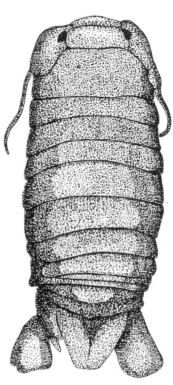

Figure 4-6. The common half-inch-long sandy beach isopod, Excirolana *sp.*

Figure 4-5. Sandy beach crustaceans known as mysid shrimp. These inch-long animals are nearly transparent in life.

small, voracious scavengers. Several different species of isopod fill this role along California's beaches; *Excirolana* spp. are the most common.

The sandy beach crustacean that's most familiar to many beachcombers, technically doesn't live in the ocean at all. This is the small amphipod crustacean known as the beach hopper or sand flea (*Megalorchestia* spp., Figure 4-7). Beach hoppers have moved out of the sea and can no longer withstand submergence in sea water. They live just above the reach of high tide where they pass the day in shallow burrows. Frequently, the high tide line contains large clumps of washed-up seaweed and other debris (called wrack), and the hoppers will burrow within or beneath it. During night-time low tides the beach hoppers emerge from their burrows and scavenge in the wrack and

along the beach for food. Hoppers prefer plant material and their favorite food is the giant kelp *Macrocystis,* but will eat almost anything organic if no plant material is available.

Beach hoppers can be found by visiting the wrack line on the upper beach and turning clumps of seaweed and other debris over. If the wrack material is fresh, it will normally harbor these animals. Small beaches or highly exposed beaches with very coarse sand sometimes will lack beach hoppers. Large beaches with extensive beach backshore and a steady

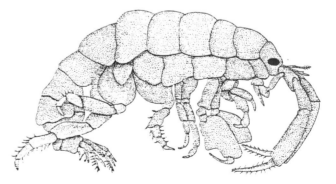

Figure 4-7. The beach hopper or sand flea, Megalorchestia *sp., a half-inch-long amphipod crustacean.*

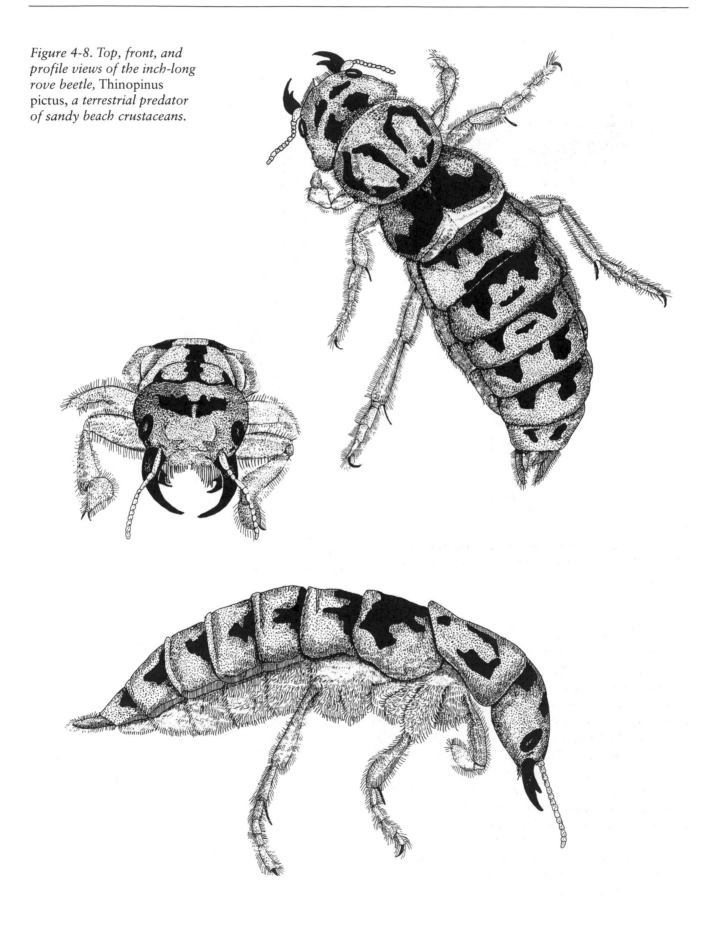

Figure 4-8. Top, front, and profile views of the inch-long rove beetle, Thinopinus pictus, *a terrestrial predator of sandy beach crustaceans.*

supply of wrack usually harbor large populations. A good time to see the beach hoppers out and about is at a low tide just after sunset or just before sunrise, when literally swarms of them can be found scavenging on the beach.

There are several species of beach hoppers on California beaches and I refer you to more technical sources to distinguish them [2, 3]. All species have a similar appearance and large hind appendages that they use to achieve the mighty leaps that give them their "hopper" name. Female beach hoppers brood their young and release them as juvenile animals right onto the beach. Species range from a third of an inch long as adults to over an inch and a half in length. Because they avoid submergence, beach hoppers are free from the normal marine sandy beach predators like fish that feed during high tide. However, it is not unusual to find large (one inch long) predatory rove beetles (Figure 4-8) in the wrack with the hoppers that prey on them. The small flies that can be so plen-

tiful in the wrack are sand flies which lay their eggs in the rotting seaweed.

Sandy Beach Polychaete Worms

There are a few obvious polychaete worms that live on the sandy beach. Usually, the most abundant on protected beaches is the blood worm *Euzonus mucronata* (Figure 4-9). The blood worm usually is found several inches to a foot deep in a distinct band that parallels the water's edge at about the mid-tide level. As the name implies, the worm is dark red and about two inches long. Blood worms can be quite abundant with dozens in a spadeful of sand. The worm burrows through the sand, swallowing it as it goes like a terrestrial earthworm. The organic particles trapped

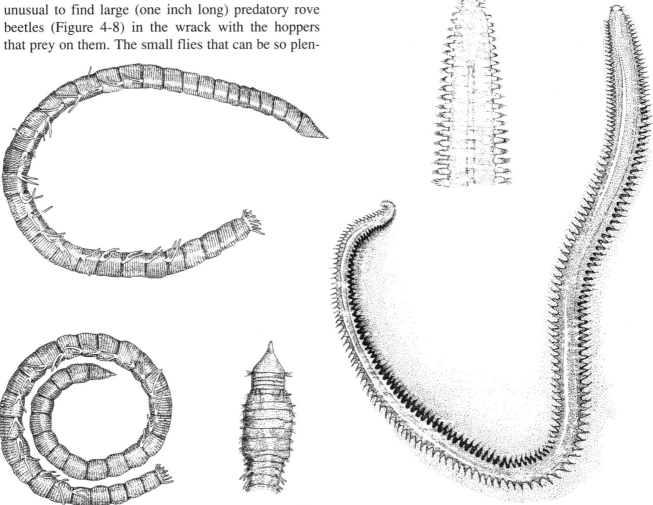

Figure 4-9. The blood worm, Euzonus mucronata, *a two-inch-long polychaete worm often abundant on sandy beaches.*

Figure 4-10. The six-inch-long shimmy worm, Nepthys californiensis, *with a close up view of the anterior.*

when the sand was deposited are digested, and the remaining sand is passed out of the gut.

When digging for mole crabs in the swash zone a shiny, gun- metal gray worm that wiggles like crazy is often discovered. This is the shimmy worm, *Nepthys californiensis* (Figure 4-10), and it occurs from British Columbia to Baja California. Shimmy worms can reach six inches in length and are predators, probably eating other worms. They are armed with a wicked feeding structure called an eversible proboscis. The proboscis is an extension of their throat with jaws at the end that they can shoot out from their mouth and snag prey. Shimmy worms also use the proboscis to help them burrow. You are in no danger from the proboscis, but just be thankful this worm isn't six feet long!

Sandy Beach Mollusks

Shelled marine snails are not abundant on sandy beaches. The combination of their locomotion across the sediment surface using a wide creeping foot and the constant wash of the waves is not a good fit. However, there are a few exceptions and these snail species burrow through the sediment instead of staying on top of it. The most common is the olive snail, *Olivella biplicata* (Figure 4-11), which occurs on sandy beaches and coastal sand flats along the entire coast line. *Olivella* has a glossy, purple high-lighted shell that reaches up to an inch and a half long. The shell is tapered at both ends, making it easy to maneuver through the sand. The very large foot has a special wedge-shaped front end that acts much like a ploughshare to allow the snail to push through the sand. The olive snail also has a long siphon that it can project up through the sand and bring water to its gills. Olives comb the sand for deposited organic material and small prey. When the tide is out, their relatively straight burrows can often be seen on the wet beach face as narrow, upraised trails with the bulge of the burrowed snail at one end. The other species of snails found on the beach are really only occasional visitors and are discussed later in this chapter.

Bivalve Mollusks. Clams are very successful filter feeders, and with all the suspended food available it is no surprise that several have adapted to the sandy beach environment. In different parts of California three species are common, and each has adapted its own unique way of living on the dynamic sandy beach. These animals are dependent on the high oxygen content of the wave-tossed water and can not survive in the quiet water of sheltered bays.

The southern most species is the bean clam or coquina, *Donax gloudi*. The bean clam is a small (to one and a half inches), wedge-shaped clam (Figure 4-12) that has very short siphons and a very large, agile digging foot. On southern California beaches as far north as Santa Barbara, bean clams can sometimes be very abundant in the swash zone, but they vary considerably in number year to year. If the clam is dislodged, its large foot quickly digs in and the clam is re-buried in a few seconds.

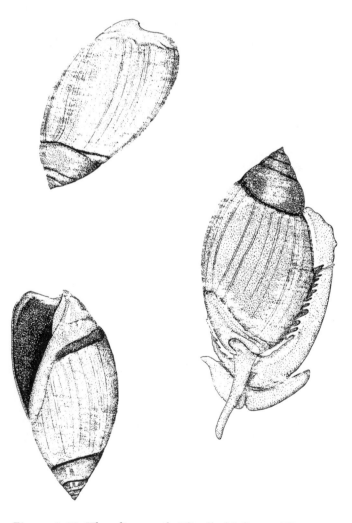

Figure 4-11. The olive snail, Olivella biplicata. *Views of the shell and extended snail showing the elongated siphon.*

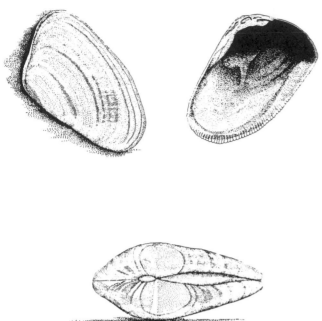

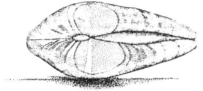

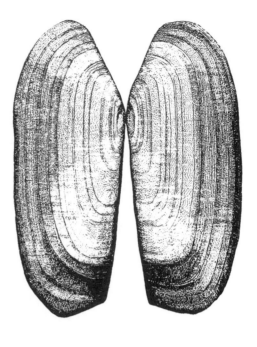

Figure 4-12. Side and top views of the bean clam, Donax gloudi, *which reaches one-and-half inches in length.*

On sandy beaches in northern California the beach-comber can find the prized razor clam, *Siliqua patula.* Razor clams (Figure 4-13) are named for their thin, elongated shells and can reach six and a half inches in length. They occur on the lowest tidal region of beaches with a flat slope and small sand grains. They often reveal themselves by a small dimple in the sand made by their siphons. It takes an experienced razor clammer armed with a narrow clam shovel to unearth one of these animals. They are very sensitive to any movement and can quickly burrow out of reach. A razor clam placed on top of the wet sand can re-burrow in ten seconds! Although known from as far south as San Louis Obispo County, razor clams are rare along the central California coast.

While the bean and razor clams can burrow like lightning, the adult Pismo clam, *Tivela stultorum,* relies on its size and its bulk to remain burrowed. The Pismo clam lives at the lowest tide mark and just beyond. The triangular shell can get quite thick and large, up to seven inches long, and weigh upwards of two pounds (Figure 4-14). Pismo clams have relatively short siphons, and so are found near the surface. The hinged side of the shell faces toward the ocean, and the broad axis of the shell is always in line

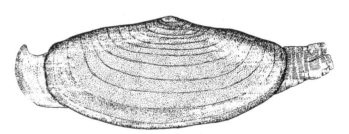

Figure 4-13. Shell and side view of the razor clam, Siliqua patula, *a rapidly-burrowing species found on northern California beaches.*

with the direction of the incoming waves to present the least resistance to the moving water. This behavior is so dependable that, during the early half of this century, it allowed the Pismo clams to be harvested with a modified hay rake pulled by horses. The clammers would determine the orientation of the clams from the direction of the wave swell and then pull the clam rake through the clam bed at right angles to this orientation. Pismo clams were scooped up by the wagon load and fed to livestock. Between 1916 and 1947, up to 50 thousand clams a year were harvested [2]. Pismo clams are not that plentiful anymore and much too delicious to feed to swine. Unfortunately, the sea otter also likes Pismo clams and can dig them

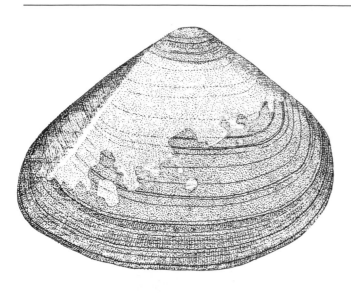

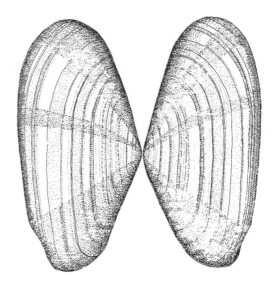

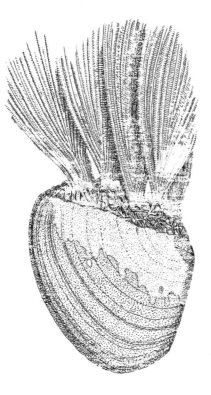

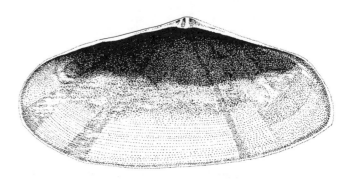

Figure 4-16. The surf clam, Tellina bodegensis, *a two-inch-long species that is common offshore of sandy beaches.*

Figure 4-14. Side view of the shell of the Pismo clam, Tivela stultorum. *The lower view shows a clam with a colony of the hyrdoid,* Clytia bakeri, *attached.*

Figure 4-15 (Right). Valves of a large (three-inch diameter) basket cockle, Clinocradium nuttallii, *a clam sometimes found on quiet sandy beaches.*

up in a hurry, sometimes wiping out a clam bed in a few weeks.

Shown attached to the Pismo clam (Figure 4-14) is a hydroid colony of the species known as *Clytia bakeri*. This colonial cnidarian isn't found on all Pismo clams, but if present is always attached on the posterior end of the shell just above the siphons, because this is the only part of the clam that is above the sand. The hydroid colony can have several two-inch branches joined by stolons (root-like attachment structures) on the clam's shell. This perch allows *Clytia* a solid purchase and access to the food-rich waters washing over the sandy beach. *Clytia* can also be occasionally found attached to the shells of *Donax gloudi* and *Olivella biplicata*.

Besides these three common beach clams, several other species may occasionally occur on the sandy beach. The beautiful basket cockle, *Clinocardium nuttallii*, which is more typical of sand flats of sheltered bays (Figure 4-15) sometimes is found on clean, fine-sand beaches. The two-inch-long surf clam, *Tellina bodegensis* (Figure 4-16), which typically lives just offshore of the sandy beach, is also occasionally stranded on the beach by strong waves.

OCCASIONAL VISITORS

In addition to the two clams previously mentioned, several other animals occasionally are found on California's beaches. I refer to these as occasional visitors because their normal habitat is either quite near or includes the intertidal beach, or they use the sandy beach to fulfill some aspect of their life cycle.

The California Grunion

Probably the most famous occasional visitor to the sandy beach is the small (six inches long), silvery fish known as the California grunion, *Leuresthes tenuis* (Figure 4-17). Grunion arrive on the beach in the middle of the night just after the peak of the highest high tides of March through August. They occur on beaches as far north as Monterey Bay, but are much more common in southern California. The fish swim up the beach with the waves and a female will be accompanied by one or more males. The female burrows into the sand backward and deposits her eggs two to three inches below the surface. As she releases

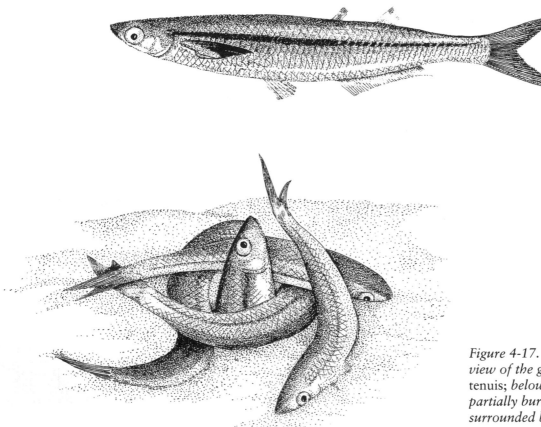

Figure 4-17. *Above: profile view of the grunion,* Leuresthes tenuis; *below: a female partially buried in the sand surrounded by males.*

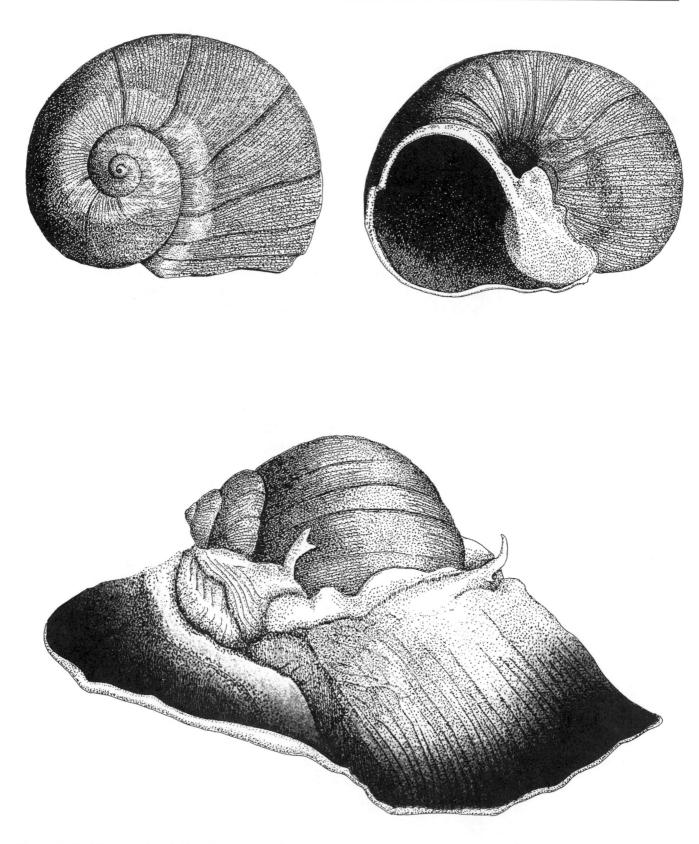

Figure 4-18. Views of the shell and expanded foot of the moon snail, Polinices *sp. Shells can reach four inches in diameter.*

her eggs, the males arch around her and release their sperm to fertilize them. Both females and males leave with the next wave. The fertilized eggs are left behind to incubate in the warm sand on the upper beach, which will not be covered with water until the next series of spring high tides. These high tides occur about ten days later, and the developing larvae burst out of their egg membranes as soon as they become wet and swim off with the next wave.

Moon Snails. A large mollusk that sometimes occurs on more protected beaches is the moon snail, *Polinices* spp. (Figure 4-18). Similar to the olive snails discussed earlier, the moon snails have a large foot with a ploughshare-like anterior modification for burrowing through sand. However, moon snail shells are very bulky (up to four inches in diameter, but usually only smaller ones are seen on the beach), and often the top third projects above the snail's burrow. This makes them susceptible to being washed out of the sand by wave action. The snails found on the beach were probably washed in from the sandy bottom just offshore.

Moon snails are more common somewhat offshore on sand bottoms and on protected coastal sand flats all along the California coast. These snails are inter-esting carnivores. They use their flexible, tongue-like radulas to bore neat, counter-sunk holes into their molluscan prey. Once the hole has been bored through the shell, the snail inserts its long snout through the opening and devours its prey. Clam and snail shells are often found on sandy beaches and coastal tidal flats with this characteristic calling card of moon snail predation (Figure 4-19).

Visiting Crabs and Shrimp. In southern California as far north as Santa Barbara the feisty swimming crab, *Portunus xantusii* (Figure 4-20), is occasionally stranded on the beach. *Portunus* comes in with the high tide to feed on polychaete worms and other beach critters, and occasionally fails to retreat when the tide ebbs. When this happens, it simply burrows into the sand and waits for the next high tide. Beach-combers may come across a crab-shaped outline on the sandy beach face and unearth a swimming crab. A word of caution about this customer, which can be up to three inches wide, *Portunus* is fast and ill tempered, and its claws can deliver a nasty nip. So handle with care or not at all! As the name implies this crab has specially modified rear legs that allow it to swim quite effectively. *Portunus* is related to the blue crab of the Eastern Seaboard, and like its close relative, it

Figure 4-19. Shell of a two-inch littleneck clam, Protothaca staminea, *showing the characteristic counter-sunk hole bored by a predatory moon snail.*

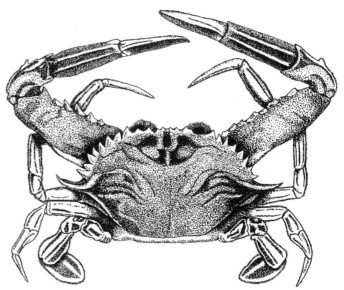

Figure 4-20. Top view of the swimming crab, Portunus xantusii. *Note the flattened, paddle-like rear legs modified for swimming.*

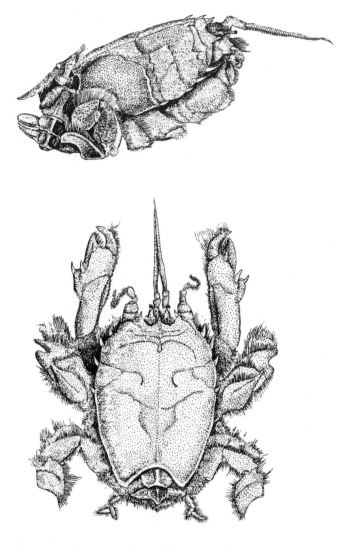

can dart up from the bottom and seize prey that swim by. It is much more commonly seen on protected sand flats of southern California coastal embayments.

Another crab occasionally found buried on the beach at low tide is the graceful cancer crab, *Cancer gracilis* (Figure 4-21). Found primarily in central and northern California, the graceful cancer is a much more docile crab than *Portunus*. *C. gracilis* can be up to four inches wide. Many of the individuals a beach-comber will find will be females carrying a large clutch of developing eggs attached to their abdomens. Like *Portunus,* the graceful cancer crab is more commonly seen on protected sand flats.

A third crab sometimes seen as far north as Point Reyes in Marin County is the spiny mole crab *Blephariopoda occidentalis* (Figure 4-22). Spiny mole crabs live just offshore of the sandy beach and can sometimes be found buried near the low water mark at low tide. They are considerably larger (up to three inches long) than their sand crab relatives, *Emerita*, and feed by scavenging food deposited on the beach.

A final crustacean seen on the beach all along California are the salt and pepper shrimps *Crangon* spp. (Figure 4-23). These two-to-three-inch-long shrimps are very active carnivores found in various shallow water habitats. It is not uncommon for a school of these shrimps to come onshore at high tide to forage for invertebrate prey, including worms and small sand crabs. If stranded, the shrimps just burrow in and wait for high tide.

Figure 4-22. Top and profile views of the three-inch-long spiny mole crab, Blephariopoda occidentalis.

Figure 4-21. Graceful cancer crab, Cancer gracilis. *Note the white-tipped claws that distinguish it from other common cancer crabs,* Cancer productus *and* Cancer antennarius.

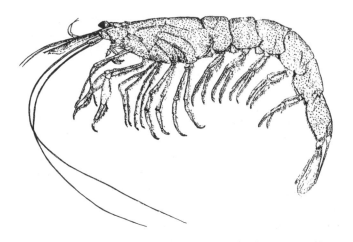

Figure 4-23. The salt and pepper shrimp, Crangon *sp., a three-inch-long predator found on sandy beaches at high tide.*

Sandy Beach Echinoderms

To the beachcomber, the Pacific sand dollar, *Dendraster excentricus*, hardly seems like an occasional visitor, because it is regularly found on most California beaches. However, what the beachcomber typically finds is actually the bleached white skeleton or "test" of *Dendraster* (Figure 4-24), which can measure up to three inches in diameter. Sometimes after a storm, living sand dollars that have been washed in from just offshore of the surf zone can be discovered on the beach. These animals are distinguished from the naked skeleton by their fine covering of short, dark spines that gives them an almost velvety appearance. The Pacific sand dollar occurs in large sand dollar beds that run parallel to the shoreline. Here the sand dollars sit upright in the sand and filter food from the water moving along shore. When storms occur and large waves are generated, the sand dollars lie flat and bury themselves in the sand, but individuals are occasionally washed out and onto the beach.

The armored sand star, *Astropecten* spp., is a predator of the sand dollars and a fellow echinoderm. Several species of these gray- colored sea stars occur along California, and all have a fairly characteristic appearance (Figure 4-25). Like their sand dollar prey, the sand stars get washed in by high storm waves and normally live offshore beyond the line of breaking waves. These sea stars can be up to a foot in diameter.

Sea Pansies

A final invertebrate living nearby that occasionally ends up on the beach is the sea pansy, *Renilla kollikeri* (Figure 4-26). The sea pansy is purple, and when inflated with water it has a flat petal about three inches across and a short stem about one and half inches long. The "flower" is actually a cnidarian colony related to the sea anemones and corals. The petal houses numerous small feeding polyps, and the stem serves to anchor the colony in the sand. Like the sand dollar and the sand star with which it co-occurs offshore, the sea pansy is occasionally uprooted and

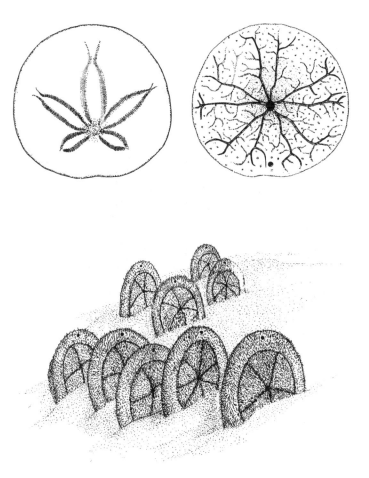

Figure 4-24. Above: the skeleton of a two-inch diameter Pacific sand dollar, Dendraster excentricus. *Below: typical feeding posture of live sand dollars.*

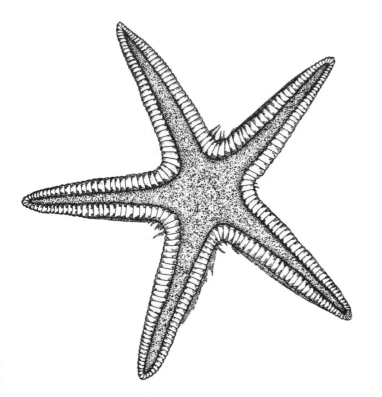

Figure 4-25. The armored sand star, Astropecten *sp., sometimes found on sandy beaches can reach up to a foot in daimeter.*

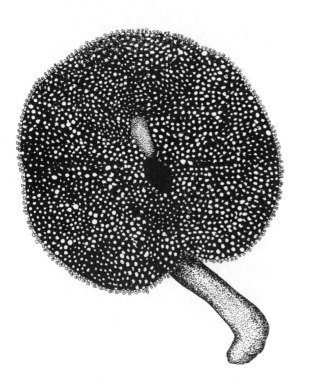

Figure 4-26. The sea pansy, Renilla kollikeri, *of southern California. This cnidarian colony can reach three inches in diameter.*

washed onshore. Sea pansies do not occur north of Point Conception, and are common animals on some southern California sandy mud flats.

WASH-UPS

The final group of organisms to consider are grouped into a category called "wash-ups." This grouping is somewhat arbitrary as many of the "occasional visitors" previously described find themselves on the beach only because they have washed up from their nearby offshore habitat. The organisms grouped here as wash-ups are animals and plants typically found well away from the sandy beach, and arrive on its surface because the relentless washes of waves have carried them in. Included here are several large planktonic animals and the large kelp plants of offshore kelp forests.

Stranded Seaweeds

Let's talk about the plants first. We've already discussed the arrival of plant material on the beach and its contribution to the "wrack" used by beach hoppers. Seaweeds can only grow attached to solid substrates where they can have access to sunlight. This limits them to rocky intertidal habitats and shallow, subtidal rocky bottoms. In rocky intertidal habitats a variety of red, green, and brown algae grows luxuriantly. Although firmly attached to the rocky substrate by root-like holdfasts, the algae can be torn loose by wave action and washed away. If a sandy beach occurs close by a rocky area, a variety of algae can be found washed up. However, if the beach is some distance from a source of algae, only the biggest, toughest algae remain intact, and the rest are shredded into detritus by wave action. Therefore, the algae most typically found on sandy beaches are the large brown algae known as kelps.

The largest of all the kelps is the giant kelp, *Macrocystis pyrifera. Macrocystis* is the predominant kelp of California's offshore kelp forests. From a stout holdfast attached to the bottom it grows vine-like branches called stipes up to 100 feet long. Broad, flat blades grow laterally from the stipes. Each blade is buoyed up by a bulbous, air-filled float located at its base (Figure 4-27). Depending on the degree of wave action and the time of the year, pieces of *Macrocystis* of differing sizes will wash up on the

Figure 4-27. The giant kelp, Macrocystis pyrifera, *showing clockwise from below: holdfast, stipe with blades, and close-up of the inch-and-half-long float.*

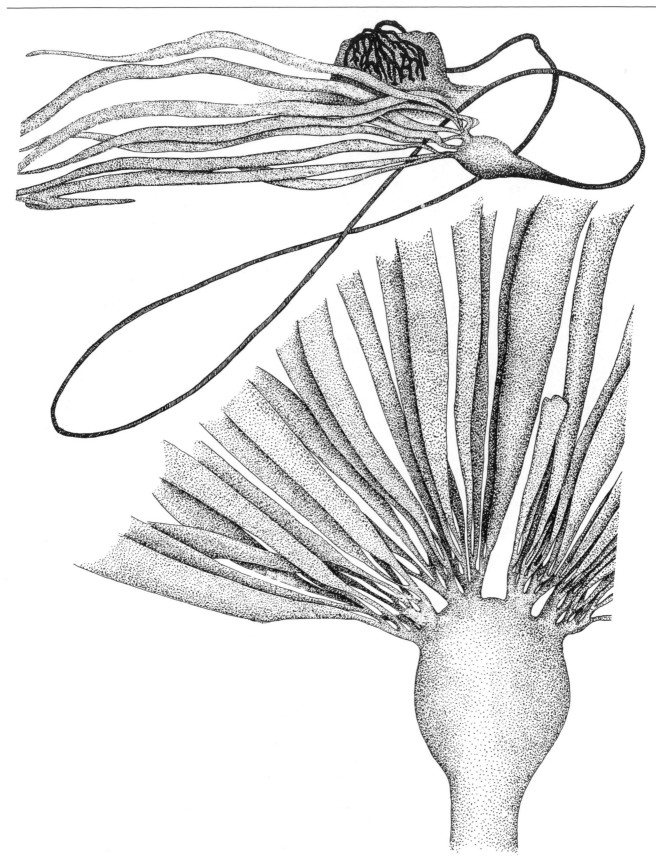

Figure 4-28. The bull kelp, Nereocystis luetkeana. *Upper: illustration of entire plant; lower: view of the single large float with blades.*

beach. During the first severe storms of early winter, whole plants will be uprooted and wash onto the beach in a huge tangled heap. Other times only a few stipes or blades will be present, and sometimes only the small inch-and-a-half-long floats will appear by themselves.

Similar in size (up to 60-70 feet long) to the giant kelp and also found in offshore kelp beds is the bull-whip or bull kelp, *Nereocystis luetkeana*. Bull kelp (Figure 4-28) differs from the giant kelp in that only a single stipe grows from a holdfast. The stipe ends in a large float from which several stout blades grow. Usually the whole plant washes up, much to the delight of junior beachcombers who find no end of uses for it. Adults on the other hand are known to slice the hollow stipes and make pickles out of them.

Two smaller kelps are often contributed from the rocky intertidal habitat. The first is the oar weed, *Laminaria* spp., which grows to about five feet long

in the lower intertidal zone. Oar weed (Figure 4-29) grows in the most wave-exposed area and takes a tremendous beating. It survives by having a very tenacious holdfast and a stipe that is very flexible and capable of bending with the back and forth wash of the waves. *Laminaria* has a stout blade that often splits as it grows.

The second intertidal species is the sea palm kelp, *Postelsia palmaeformis* (Figure 4-30). The sea palm grows to about three feet high in the middle intertidal zone of the most exposed rocky intertidal habitats. Like the oar weed, the sea palm weathers the onslaught of the waves by having a flexible stipe topped with short, palm-like blades. Both the oar weed and the sea palm can be found on sandy beaches. The sea palm often occurs in small clumps of several stipes attached to a common holdfast.

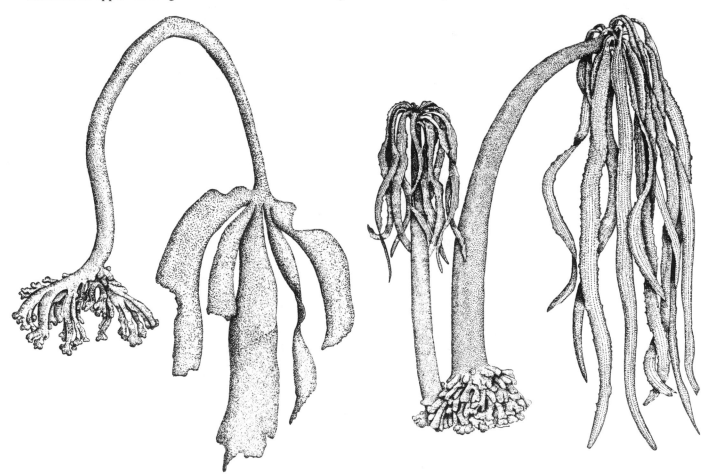

Figure 4-29. The oar weed, Laminaria *sp., a kelp plant that washes in from the low rocky intertidal zone.*

Figure 4-30. The sea palm kelp, Postelsia palmaeformis, *showing two plants attached to a common holdfast.*

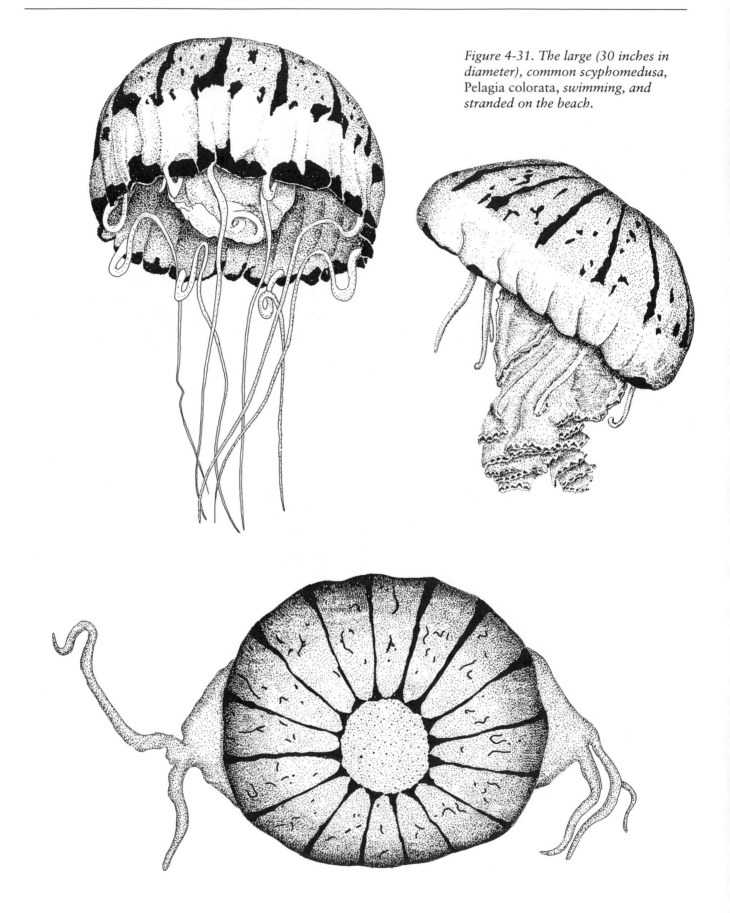

Figure 4-31. The large (30 inches in diameter), common scyphomedusa, Pelagia colorata, *swimming, and stranded on the beach.*

Stranded Jellies

The broad, exposed sandy beaches collect an amazing array of gelatinous marine zooplankton. These species vary from small, marble-sized ctenophores to medusas over a yard in diameter. Usually when these animals are found by the beachcombers they are unrecognizable mounds of jelly. However, if they are newly arrived, they can be quite intact and very interesting to investigate. The "jelly" is a special type of animal material, called mesoglea, that allows the organism to float. Mesoglea is mostly water with a chemical composition slightly different from sea water that makes it a little bit lighter. Although most of these gelatinous zooplankters have some means of locomotion, they can't out swim the force of the currents. If the water mass in which they're floating comes too close to shore, they can become stranded, sometimes in very large numbers.

Scyphomedusas. The large medusas (greater than a foot in diameter) that wash up all belong to the class Scyphozoa of the phylum Cnidaria. Cnidarians all possess special cell capsules known as nematocysts, which they use to stun and entangle prey. These large medusas, known as scyphomedusas, have nematocysts all over their bodies not just on their tentacles. Even though a medusa stranded on the beach appears motionless and apparently dead, the nematocysts are still potentially capable of discharging and imparting a nasty sting to an unsuspecting beachcomber. So to be safe, don't handle any large stranded jellyfish with your bare hands. If you wish to investigate them further, use gloves or a stick to turn them.

When the medusa is seen swimming in the water column, it is a magnificent animal to behold (Figure 4-31). The bell or umbrella contracts rhythmically, and the animal is propelled gracefully upwards with its long tentacles and oral arms extended out below it to entrap prey. However, when it becomes stranded on the beach, it is robbed of the water's supporting buoyancy and appears as a round, flat, motionless blob of jelly.

The most striking of the commonly stranded scyphomedusas is the purple-striped *Pelagia colorata* (Figure 4-31), that can reach a bell diameter of over 30 inches. It has eight long tentacles spaced evenly around the bell margin, and oral arms several yards in length extending from the mouth in the center of the bell. Nematocysts of this species sting fiercely, so be

very careful with them. *Pelagia* occur all along the continental shelf of California, and sometimes large numbers will be trapped in coastal embayments like Elkhorn Slough on Monterey Bay.

Another medusa commonly stranded along the entire California coast is the sea nettle, *Chrysaora melanaster* (Figure 4-32). The bell, which can reach a diameter of over 15 inches, has radial yellow or brown lines and 24 reddish marginal tentacles. The frilly oral arms of the sea nettle are quite prominent.

The moon jelly, *Aurelia aurita*, has a world-wide distribution. It sometimes occurs in large nearshore aggregations along California, resulting in many strandings. *Aurelia* reaches a bell diameter of 15 inches and has numerous short tentacles around the margin of its scalloped, saucer-shaped bell (Figure 4-33). The bell is translucent gray or blue with four

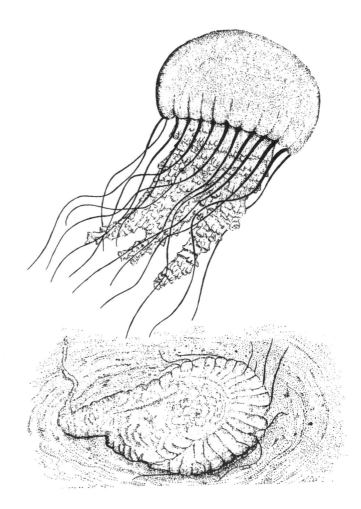

Figure 4-32. The 15-inch diameter sea nettle, Chrysaora melanaster, *as viewed in the water and stranded on the beach.*

Hydromedusas. Many smaller medusas also become stranded along California's beaches. These animals belong to the cnidarian class Hydrozoa, and are called hydromedusas. They usually have a sessile polyp form in their life cycles as well. Only two of the largest, most common species are included here. Further information on hydromedusas can be found in more detailed references [3, 4].

The first species is the graceful *Polyorchis pencillatus* (Figure 4-34). *Polyorchis* can be very common in coastal embayments in the spring, and often large numbers can be found stranded along the adjoining beaches. *Polyorchis* is about the size of a hen's egg with one end opened outward to form the bell margin. The rather tall umbrella is ringed with up to 90 tentacles each with a bright red eyespot at its base where it joins the bell margin. *Polyorchis* has four prominent radial canals each with numerous long, slender gonads

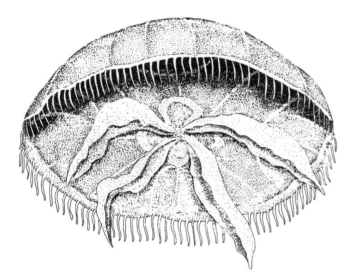

Figure 4-33. Top and bottom views of the common moon jelly, Aurelia aurita, *which can reach 15 inches in diameter.*

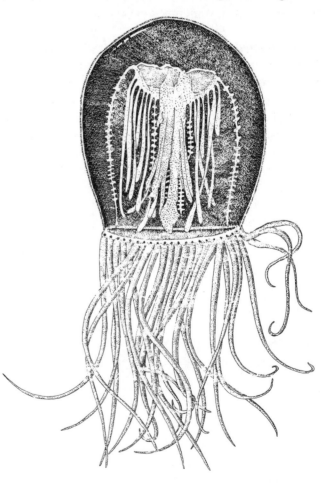

Figure 4-34. The egg-sized hydromedusa, Polyorchis pencillatus. *Note the eyespots at the base of the tentacles.*

large, horseshoe-shaped gonads clearly visible through the top. The male gonads are purple and those of the females are yellow to white. The moon jelly is unique among the large medusa species in that it does not capture its prey with its tentacles. Instead it traps zooplankton in mucus coating the outer surface of the bell and the oral arms. The sessile polyp stage of *Aurelia* is well known, and can occur in large numbers attached to floats in quiet harbors.

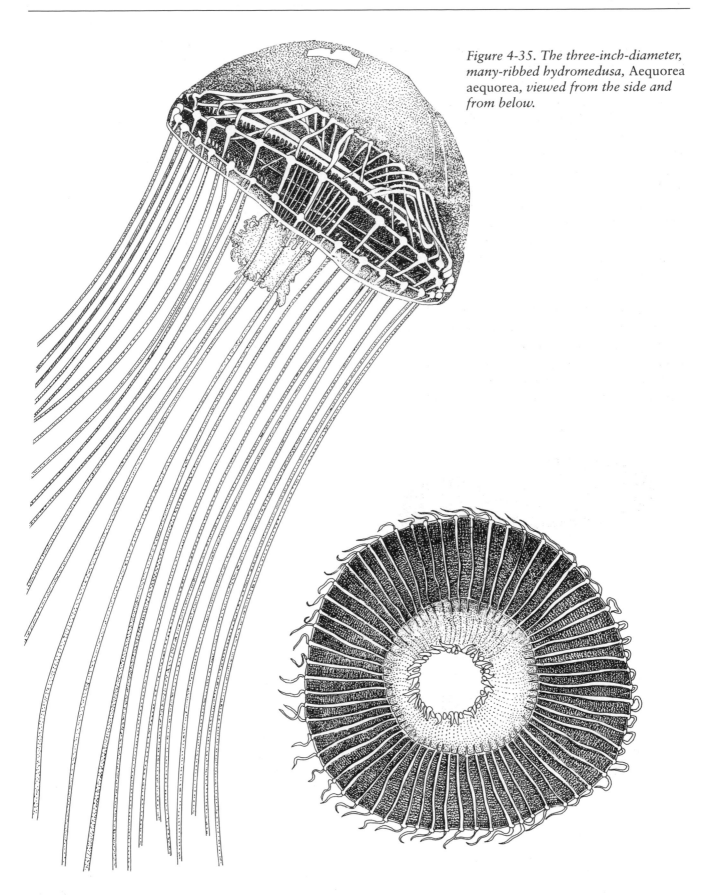

Figure 4-35. The three-inch-diameter, many-ribbed hydromedusa, Aequorea aequorea, *viewed from the side and from below.*

suspended from them hanging down into the bell. *Polyorchis* has relatively harmless nematocysts and is small enough so that a freshly stranded specimen can be suspended in a plastic bag of sea water. Try it!

While *Polyorchis* has a tall graceful bell, the many-ribbed hydromedusa, *Aequorea aequorea* (Figure 4-35), is saucer-shaped. *Aequorea*'s bell reaches over three inches in diameter and is about an inch in height. The bell is thick and glossily transparent with 60 narrow, radial canals and up to 80 long, marginal tentacles. The mouth is surrounded by ruffled lips. *Aequorea* is very bioluminescent. At night, live stranded specimens can give off a bright flash of light when disturbed.

By-the-Wind Sailors. Another cnidarian "jellyfish" that can be very abundant on California's beaches is not a medusa at all, but instead is a large (up to three inches long) floating polyp. This is the by-the-wind sailor, *Velella velella* (Figure 4-36). *Velella* has a stiff exoskeleton made of chitin, which includes a gas-

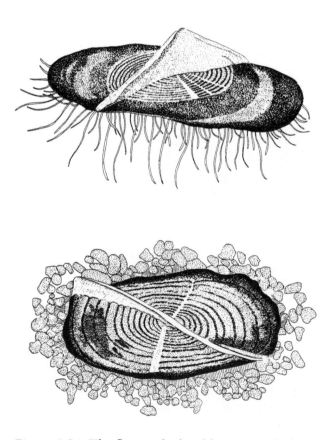

Figure 4-36. The floating hydroid known as the by-the-wind sailor, Velella velella, *which can reach three inches in length.*

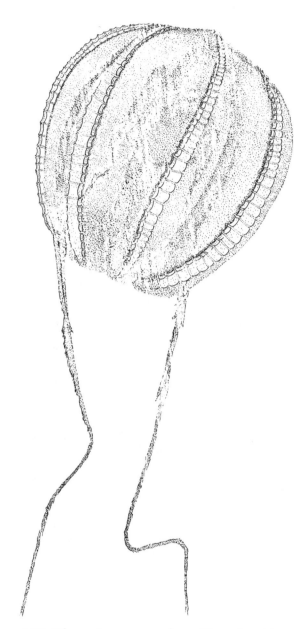

Figure 4-37. The cat's eye ctenophore, Pleurobrachia bachei, *swimming in the water. Stranded animals are the size of a marble.*

filled float and a transparent sail. The sail is mounted diagonally to the left or right of the long axis of the body, so that the animal tacks at a 45° angle to the left or right of the true wind direction. *Velella* with right-angled sails predominate off California, and they remain offshore as long as the wind is from the north. However, when the wind shifts to westerly or southerly, the animals are driven onshore. In central California the wind pattern shifts to a northwesterly direction in early spring and, depending on its

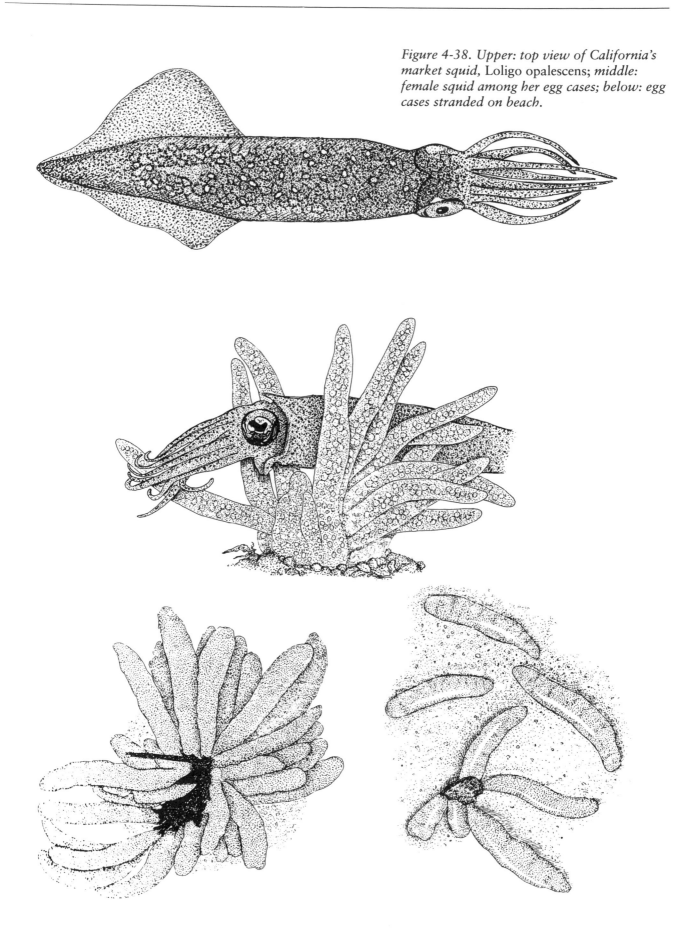

Figure 4-38. Upper: top view of California's market squid, Loligo opalescens; *middle: female squid among her egg cases; below: egg cases stranded on beach.*

strength and persistence and the abundance of *Velella* offshore, windrows of beached by-the-wind sailors several inches deep can accumulate on exposed sandy beaches. These animals create quite a stink for a short time, but quickly decompose, leaving behind only the thin, blue, cellulose-like chitinous exoskeletons which can persist on the beach for months.

Comb Jellies. Sometimes the beach can be littered with round, marble-sized jellies. Often these are also not medusas, but closely related animals known as ctenophores or comb jellies. The most common animal is the sea gooseberry or cat's eye ctenophore, *Pleurobrachia bachei* (Figure 4-37). Ctenophores lack the stinging nematocysts typical of cnidarians, and instead use tentacles with special adhesive cells to snare their zooplankton prey. Ctenophores can be quite marvelous to behold when suspended in a plastic bag of sea water. Their eight rows of ciliary plates (ctenes) beat rhythmically and catch the light in a flashing spectrum of color

Central California beachcombers have the unique opportunity to view many of these jellies at the marvelous jellyfish exhibit at the Monterey Bay Aquarium in Pacific Grove, California. The jellyfish are maintained in specially designed aquaria that allow them to swim freely for the public to view through viewing ports. The jellyfish display alone is worth the price of admission, and there is so much more to see at this flagship of the marine aquaria.

Squid Egg Cases. Sometimes in summer or early fall, California beaches may be strewn with a few or many, six-to-eight-inch-long, quarter-to-half-inch-in-diameter jelly sausages. Closer inspection reveals that the sausages themselves are packed with round, quarter-inch-in-diameter spheres. These are the egg cases of California's market squid, *Loligo opalescens* (Figure 4-38). The squids mate in large aggregations some distance offshore over sandy bottoms. The females receive special sperm packets from the males, fertilize their eggs, and then extrude the eggs into these special egg cases that they anchor into the sandy bottom. Often, either by wave action or some other disturbance, the cases are uprooted and wash ashore. Depending on the degree of development the embryos have achieved, tiny squid may be seen through the egg membranes with the use of a hand lens.

Attached Pelagic Animals

Pelagic Barnacles. Often after a particularly severe storm, the more exposed beaches will be littered with pieces of debris that have been floating at sea for sometime. These can include such mundane junk as bottles and plastic bags or such exotica as glass or plastic fishing floats used by the Japanese on nets in their mid-Pacific fisheries. These floating substrates are the homes of a sea-going group of crustaceans known as pelagic stalked barnacles. Barnacles make their living by filtering food out of the water using the hairs on their body appendages. The closer together the hairs, the smaller the particles they can filter. On central and northern California beaches large pieces of driftwood are sometimes covered with the barnacle *Lepas anatifera* (Figure 4-39). This species is unmistakable from other pelagic stalked barnacles because of its large size (up to six inches long) and the smooth white plates that make up its shell. It is also found on glass floats, bottles, and styrofoam.

A smaller (one inch long) stalked barnacle, *Lepas pacifica* , is commonly stranded by seasonal onshore winds in southern California. This barnacle can be found on wood, seaweed, feathers and a host of other, smaller floating debris including the exoskeleton of by-the-wind sailors. It has thin shells that show the blue color of the underlying tissues.

Figure 4-39. Five-inch-long pelagic stalked barnacles, Lepas anatifera, *attached to a piece of floating wood.*

Shipworms. Driftwood that has been at sea for sometime is often riddled with wood-boring invertebrates. If the wood is soft enough, breaking off a piece may reveal the white, four-fifth's inch diameter, calcium-carbonate lined burrows of shipworms. Shipworms are not worms at all, but are bivalve mollusks with shells specially modified for boring into wood. Their fleshy bodies can reach over a yard in length and are tipped by siphons that reach to the burrow opening (Figure 4-40). The siphons are tipped with calcareous, feathery-looking structures known as pallets that keep the opening intact.

The Dungeness Crab. A final animal found on central and northern California beaches in the late winter really isn't there at all! No, it is not a ghost, but the molted outer skeleton (exoskeleton) of the large male, market or Dungeness crab, *Cancer magister* (Figure 4-41). In the late winter male Dungeness crabs molt offshore in close temporal synchrony in preparation for the spring and summer breeding season. They can be so abundant in some years that large windrows of cast exoskeletons pile up on sandy beaches (Color Plate 13a). Because the crab molts its entire outer covering intact, a fresh molt can easily be mistaken for a live animal. However, as the molt washes about it pulls apart and soon is in pieces. The solid, shield-like carapace that covers the head and body of the crab remains intact for sometime, and is often found by itself on the beach.

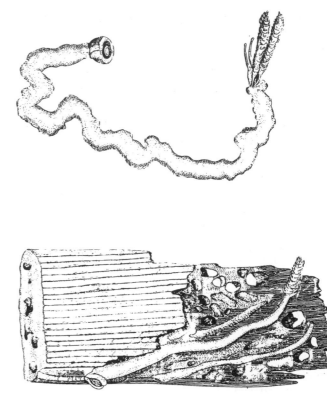

Figure 4-40. Below: driftwood riddled with the burrows of boring bivalve mollusks known as shipworms; above: a clam removed from its burrow.

Figure 4-41. A seven-inch-wide male Dungeness crab, Cancer magister. *This crab is light brown to buff-colored in life.*

A FINAL NOTE ON SANDY BEACHES

This is only a list of the very common organisms that typically show up on California's sandy beaches. Because of the immense reaches of shoreline the beaches traverse, virtually anything in the sea may show up at some time or another. A 60-foot-long blue whale washed up 15 miles from my home one summer! Many of the organisms presented elsewhere in this guide may wash up on the beach, so flip through the other chapters if you are stumped. Sometimes knowing what the adjacent shoreline looks like can be helpful. If there is a rocky intertidal habitat nearby, many of the shells and algae probably came from there. If the beach is near the outlet of an estuary or enclosed embayment, the organisms may have been carried out with the tide and washed ashore on the beach. Keep your eyes open and good hunting!

References

1. Bascom, W. *Waves and Beaches.* (revised), Garden City: Anchor Press, 1980, 366 pp.
2. Ricketts, E. et al. *Between Pacific Tides,* 5th ed. Stanford:Stanford University Press, 1985, 652 pp.
3. Smith, R.I. & J. Carlton. *Light's Manual: Intertidal Invertebrates of the Central California Coast,* 3rd ed. Berkeley:University of California Press, 1975, 716 pp.
4. Kozloff, E.N. *Seashore Life of the Northern Pacific Coast.* Seattle: University of Washington Press, 1983, 370 pp.

5
Quiet-Water Habitats: Bays, Estuaries, and Lagoons

HOW QUIET-WATER HABITATS ARE FORMED

The California coastline is constantly changing. Sandy beaches can be transformed in a matter of hours during winter storms. Coastal bluffs are continuously weathered by wind and wave. Sea level changes and plate tectonic activity over geologic time have alternately inundated and uncovered vast areas of coastal land. Against this backdrop of change, a number of habitats are created that have one thing in common. These are coastal areas that somehow become shielded from the onslaught of the waves, and as a result develop into quiet-water habitats.

Estuaries are examples of such habitats. They are typically formed when a large river is flooded by rising sea level. The sediment carried from the land accumulates as a delta in the drowned river mouth. Sand flats, mud flats, and finally salt marshes develop as the delta grows and stabilizes. Another example would be a strong longshore current transporting sediment along the coastline and depositing it behind a headland. The deposited sediment builds up a sand bar and gradually cuts off a cove behind the headland, forming a protected coastal embayment. As tidal action moves water in and out of the embayment, the fine sediments are deposited, and sand and mud flats are formed. A final example would be a coastal lagoon. Sea level drops and rivers carve large channels into the exposed, wave-cut terrace. Sea level again rises and the channels become flooded with salt water. Subsequently, the connection to the sea becomes constricted or closes, and a trapped body of sea water, i.e., a lagoon, is created.

Quiet-Water Marine Environments

These quiet-water, soft-bottom habitats provide an environment very different from the sandy beach. Here organisms can create burrows in the soft substrate that will be somewhat permanent. An organism can move across the substrate at high tide and not be washed away by wave action. Tidal action continues to bring suspended food into these habitats that combines with local plankton production to provide ample food for a host of filter feeders. Finally, as the water becomes quiet in the most sheltered regions of these habitats, the fine organic material carried in suspension is deposited. This provides food for animals that feed on organic matter that accumulates on and within the soft sediment.

Because the types of substrates and opportunities for feeding are similar in these quiet-water habitats, they tend to host similar types of animals. However, there are some differences. Coastal embayments typically contain normal, undiluted sea water. In comparison, estuaries have a gradient of salinities running from fresh water at the head of the estuary, where the river or other fresh-water source enters, to pure sea water at the mouth of the estuary where it enters the sea. Salt-water lagoons vary in salinity depending on the source(s) of water that supplies them.

Estuarine Organisms. The internal body fluids of marine animals contain essentially the same types of chemical elements (e.g., sodium, chloride, calcium, etc.) as pure sea water, and in the same concentrations. When the salinity of the water varies, these animals become stressed. Thus a lagoon, a coastal embayment, and the more saline water at the mouth of an estuary all can potentially support a very similar marine fauna. However, the upper, less saline reaches of the estuary will harbor a unique fauna adapted to the lower salinities that occur there.

A final word on estuarine organisms. Very few marine animal groups have successfully evolved the ability to withstand the low salinities found in estuaries. On the other hand, those groups that have adapted to estuarine conditions tend to be very hardy. They are easily transported by human conveyances such as in the ballast water tanks of ships or among estuarine food species, like oysters, transported for human consumption. The estuarine fauna along the California coast is very poorly developed, and the native fauna found in a given estuary contains very few species. Because California's estuaries have such limited faunas, and because estuaries, such as San Francisco Bay and Humboldt Bay, tend to be such intense sites of international human commerce, large numbers of estuarine animals from around the world have been introduced there.

Shorebirds. It will be obvious to beachcombers that the most conspicuous animals often seen in these quiet-water habitats are the shorebirds that forage there during low tide, and other water fowl that take refuge on the clam water during high tide. Salt marshes are great places to view herons and snow-white egrets. Mud flats team with foraging shorebirds with every combination of bill and leg length. Unfortunately, there is not room to cover birds in this guide. There are several excellent guide books to western birds [1,2] and marine birds [3,4,5], and I refer you to them.

In this chapter I will discuss specific quiet-water habitats such as sand flats and mud flats and introduce the marine organisms that are found commonly in these habitats wherever they occur. I will also point out those organisms that are unique to these habitats in estuaries.

MUD FLATS

The Mud Flat Environment

In the quietest reaches of protected habitats, the water becomes very still, and the finest particles carried in suspension by the water are able to settle out.

These fine sediments are chiefly clay and slit particles and collectively they form muds. A word of caution to beachcombers about mud flats. They can be so soft that they will not support a person's weight, and you can quickly sink up to your boot tops and become mired. Always proceed with caution onto any soft sediment substrate. Use a shovel or some other probe to determine how soft the sediment is ahead of you. Depending on the composition of the substrate, you may be able to continue with only a slight bit of sinking or you may decide to move somewhere else and try again.

It should be pointed out here that although mud flats and sand flats are treated as separate habitats, they are part of a continuum of sediment types that grade from the finest clay muds to cobble stone beaches on up. In the quiet-water habitats discussed here the soft substrate is often a mixture of several sediment sizes reflecting the different types of water movement that occur over a particular area. For example, a region in an embayment that is generally only influenced by gentle tidal currents may occasionally experience locally generated waves when the wind comes from a different direction than normal. This could result in some larger sand grains being transported in and mixing in with the finer sediments. Marine biologists might describe one such sediment as a sandy mud and another, with slightly more sand, as a muddy sand, i.e., arbitrary terms that suggest the general composition of the substrate. Therefore, although this section is titled mud flats, many of the organisms discussed are found over a range of soft sediment habitats.

Because of the gradual way mud flats are formed, they tend to be quite flat with little physical relief. However, the surface signs of the burrowing and feeding activities of the resident organisms can sometimes be quite extensive (Figure 5-1). Mud flats might seem to be the last place you'd find seaweeds, but sometimes extensive growths of the green algae *Ulva* spp. and *Enteromorpha* spp. (Color Plates 1a and 1b) can be found. Also, occasional red algae will grow attached to an exposed shell or other debris. There are even some species of red algae that specialize in growing in sandy mud substrates. Often, mud flats will be fringed by salt-marsh vegetation, and this habitat is discussed later in this chapter.

(text continued on page 65)

Plate 1a. The sea lettuce, Ulva sp.

Plate 1b. Enteromorpha *sp.,growing on a cliff face.*

Plate 1c. Pin cushion alga, Cladophora.

Plate 1d. Dead man's finger alga, Codium fragile.

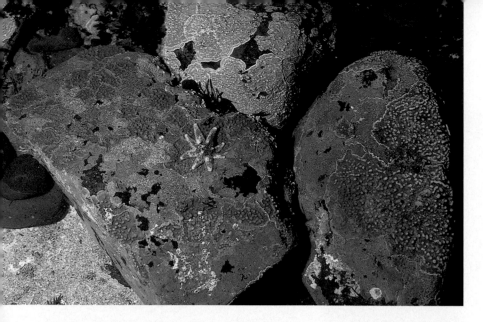

Plate 2a. Rocks covered with encrusting coralline algae.

Plate 2b. Erect or articulated coralline algae.

Plate 2d. Two red algae, Endocladia muricata (wiry alga), and Gigartina sp.

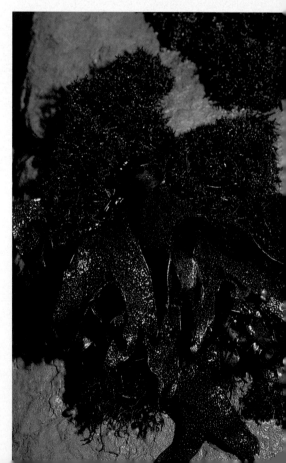

Plate 2c. Porphyra sp., a red alga commonly known as laver or nori.

Plate 3a. The rockweeds Fucus *sp.*
(larger alga) and Pelvetia sp.

Plate 3b. Surf grass, Phyllospadix *sp., with red algal epiphyte,* Smithora naiadum.

Plate 3c. Eel grass, Zostera marina,
growing on an estuarine sand flat.

Plate 3d. Red algae of the middle to low intertidal zones, Iridaea spp.

Plate 4a. Tube-dwelling anemone, Pachycerianthus fimbriatus.

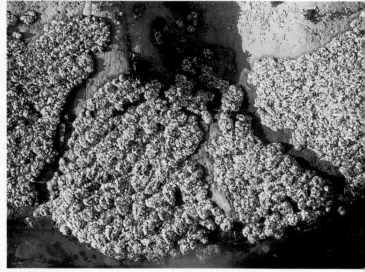

Plate 4b. Moonglow anemone, Anthopleura artemisia.

Plate 4c. Three clones of the anemone Anthopleura elegantissima; *open rock spaces separate the clones.*

Plate 4d. Large, solitary individuals of Anthopleura elegantissima.

Plate 5a. Giant green sea anemone, Anthopleura xanthogrammica.

Plate 5b. The proliferating anemone, Epiactis prolifera. *Note young on column.*

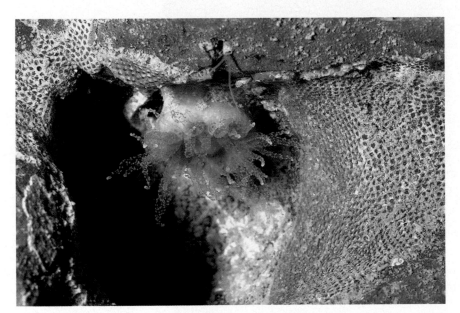

Plate 5c. Solitary orange cup coral, Balanophyllia elegans.

Plate 5d. Two clones of the strawberry anemone, Corynactis californica.

Plate 6a. Low intertidal and subtidal anemone, Tealia lofotensis.

Plate 6b. Tubes of the phoronid worm, Phoronopsis viridis, *exposed by wave action.*

Plate 6d. The nemertean Paranemertes peregrina, *shown capturing an isopod.*

Plate 6c. The nemertean worm, Tubulanus polymorphus, *near a purple sea urchin.*

Plate 7a. Sand flat polychaete worms, Lumbrinereis zonata, exposed in a shovel full of sand.

Plate 7b. Hairy-gilled worm, Cirriformia spirabrancha.

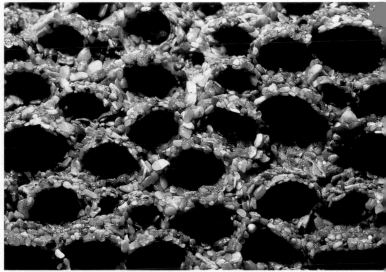

Plate 7c. Close up of the tubes of the sand castle polychaete worm, Phragmatopoma californica.

Plate 7d. Red and white filter-feeding tentacles of the fan worm, Serpula vermicularis, expanded above its calcium carbonate tube.

Plate 8a. The mossy chiton, Mopalia lignosa.

Plate 8b. The chiton, Stenoplax heathiana.

Plate 8d. Mertens' chiton, Lepidozona mertensii.

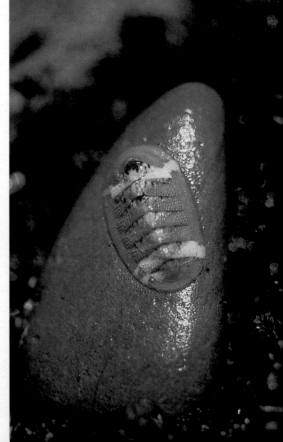

Plate 8c. The lined chiton, Tonicella lineata.

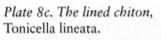

(text continued from page 56)

Mud Flat Cnidarians. Soft sediment habitats may seem like strange places to find cnidarians because their nematocyst-bearing tentacles give them a typically delicate appearance. However, cnidarians are very simple animals that breathe across their body surface and have no elaborate gills to clog with sediment. Therefore, as long as they have access to oxygenated water, they can survive quite well in soft substrate habitats.

On Newport-San Diego area mud flats the beachcomber might come across large forests of the large fairy palm hydroid, *Corymorpha palma* (Figure 5-2). This cnidarian has individual flower-like polyps (hydranths) mounted on stalks up to five inches long. The stalks are anchored in the mud by root-like processes. When the tide is out, the fairy palm hydroid droops over and looks watery and transparent. However, when the tide returns the hydroid becomes upright and extends its tentacles for feeding.

Another large cnidarian of these southern mud flats is the elongated sea pen, *Stylatula elongata* (Figure 5-3). This greenish-gray, colonial animal is a close relative of corals and sea anemones. Along the stiff, 24-inch-long central axis are numerous small polyps mounted on short leaflike flanges. The sea pen can move rapidly up and down its vertical burrow, and is buried almost out of sight at low tide. At high tide the sea pens emerge and can occur in large numbers

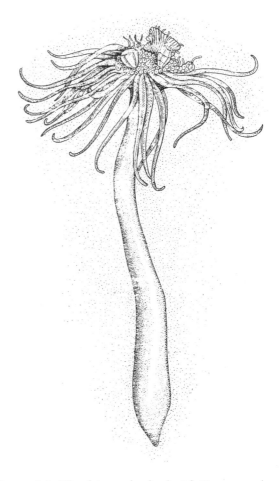

Figure 5-2. The fairy palm hydroid, Corymorpha palma. *This cnidarian can reach five inches in height.*

Figure 5-1. The surface of a mudflat in San Francisco Bay, disturbed by the activities of the burrowed organisms.

Figure 5-3. The elongated sea pen, Stylatula elongata; *left: upper portion of animal as it would appear at high tide; right: entire animal.*

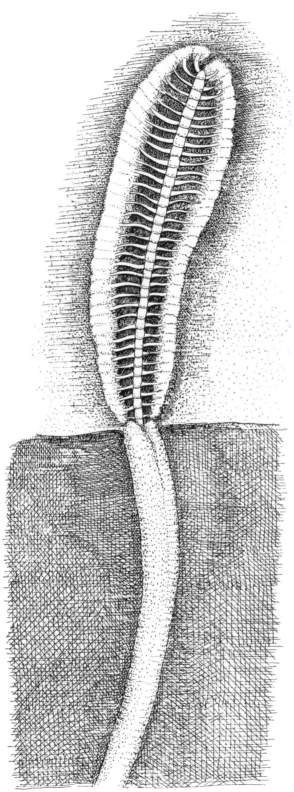

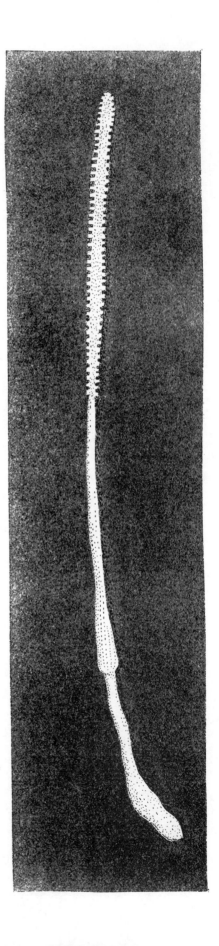

below the mid-tide level on sandy mud flats. Farther north they only occur offshore on soft bottoms to depths of over 200 feet.

A third cnidarian of southern California mud flats is a tube-dwelling, anemone-like animal known as *Pachycerianthus fimbriatus* (Color Plate 4a). This animal can reach 14 inches long when fully extended and lives in a black, slippery tube that projects above the mud bottom. *Pachycerianthus* emerges from the tube at high tide and spreads a double whorl of tentacles to collect zooplankton prey. If disturbed, the cnidarian can quickly retreat into the safety of its tube.

Several other small cnidarians inhabit mud flats. Small anemones, typically with bulbous, mushroom-like anchors, emerge on the sediment surface at high tide and spread their tentacles for feeding [6].

Polychaete Worms. The mud flat is a haven for a bewildering variety of worms. The polychaete worms are particularly well represented here. It is impossible to cover all the species a beachcomber might find by digging into the substrate. Therefore, only the largest, most abundant worms are included in this guide.

The lugworms (Figure 5-4) of the genera *Arenicola* and *Abarenicola* give their presence away by the formation of mounds of fecal material on the sediment surface. A lugworm species can be up to 12 inches long, and lives in an L-shaped burrow with its head facing into the base of the L. It uses its blunt proboscis to engulf the sandy mud sediment that continues to fall into the burrow from the surface, forming a recognizable dimple. The worm passes the ingested sediment through its digestive tract to remove any organic matter contained, and the remainder is defecated onto the surface in the characteristic fecal mounds.

Another large polychaete worm that announces its presence on the sandy mud flat is *Pista pacifica* (Figure 5-5). Pista lives in a tube formed of stiff mucus mixed with sand. The tube projects an inch or so from the surface as a fringed hood as seen in the illustration. During high tide the worm moves to the top of its tube and spreads out a wreath of long tentacles along the sediment surface. The tentacles are quite sensitive and pick up deposited organic matter and transport it back to the worm's mouth located in the center of the tentacles. The beachcomber may find quite a few of these funnel-shaped hoods on a mud or sand flat, but never see the worm. The tube of a large

worm (they are up to 15 inches in length) may reach three feet into the sediment. Before you can reach the worm at the bottom of its tube, the sides of your excavation cave in around you. Take my word for it, they're under there!

Sometimes a shovel full of sandy mud will reveal a large (up to 14 inches long), red polychaete with a pointed snout. This is a bloodworm, *Glycera* sp. (Figure 5-6). Several species occur along California, the more common being *Glycera robusta* illustrated here, and *Glycera americana*. These worms are voracious carnivores. They excavate a series of interconnected burrows called a gallery and wait for prey to pass over and disturb the water in the gallery burrows. When an unsuspecting worm or small crustacean wanders over the gallery, *Glycera* quickly emerges and everts a proboscis that can stretch over a third the worm's length. The proboscis ends in four black, hooked jaws each equipped with its own poison gland. Prey items are quickly subdued and swallowed whole. If you should dig up a bloodworm, don't pick it up. The larger specimens can deliver a nasty bite.

Another large errant polychaete found in sandy mud is the clam worm, *Neanthes brandti* (Figure 5-7). The clam worm can reach up to a yard in length and over an inch in circumference. It is a nereid polychaete, as described in Chapter 3, and occurs in a variety of habitats including sandy mud flats and the rocky intertidal zone. Clam worms are omnivorous, taking whatever comes along including smaller

Figure 5-4. A lugworm, Arenicola *sp., a burrow-dwelling polychaete. Note the blunt anterior and tapered posterior of this four-inch-long specimen.*

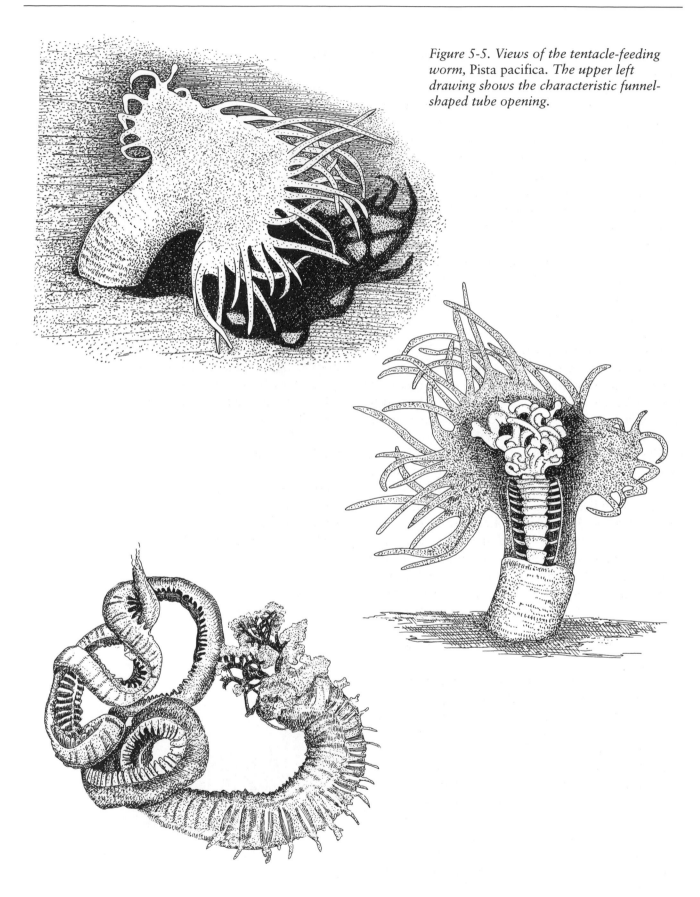

Figure 5-5. Views of the tentacle-feeding worm, Pista pacifica. The upper left drawing shows the characteristic funnel-shaped tube opening.

worms and bits of algae. Several other, smaller nereid species may be found in the soft sediment habitats. They are all recognizable by the familiar set of nereid head appendages as seen here in the clam worm.

An excavation in a sandy mud flat may often turn up a tangle of thin, red worms. They can be identified as polychaetes because close inspection reveals that their stretched bodies are segmented. A variety of polychaete species burrows through the sediment ingesting it as they go like earthworms. Illustrated here is *Lumbrinereis zonata* (Color Plate 7a), which can reach up to three feet long when fully extended. These worms can be very abundant in Tomales Bay in Marin County, and Elkhorn Slough in Monterey County, and are found on mud flats from Alaska to Baja California.

Figure 5-7. *Anterior of a clam worm,* Neanthes brandti, *a large polychaete found in several California habitats.*

What appears to be a jumble of small, dark-green worms on the mud flat surface may actually be a single worm, the hairy-gilled polychaete, *Cirriformia spirabrancha* (Color Plate 7b). When dug up, the "worms" end up being the feeding and breathing tentacles of a five-inch-long, yellow to green worm that remains buried safely beneath the surface. This polychaete is also very abundant in Tomales Bay and Elkhorn Slough and are found on mud and sandy mud flats from central California to Baja California.

A final polychaete to look for is *Hesperone adventor* (Figure 5-8). *Hesperone* has a series of flattened plates or scales that cover its back, which give it its common name, scale worm. This worm reaches two inches in length and lives in the same burrow as the large innkeeper worm, *Urechis caupo* and feeds on its discards. Several other similar-appearing scale worms occur in the burrows and tubes of large invertebrates [7].

Other Mud Flat Worms. *Urechis caupo* (Figure 5-8) is an echiuroid worm that can reach a length of 20 inches. It lives in U-shaped burrows on sandy mud flats at mid tide level or below all along the California coast. The two openings often are marked by low, volcano-shaped mounds of sediment. *Urechis* is a filter feeder. It spins a fine-meshed mucous net that it attaches to one end of its burrow. It then pumps water through its burrow and net using peristaltic contractions of its body. Periodically the innkeeper ceases contractions and ingests the net. Particles that are discarded are fought over by the scale worm *Hesperone*

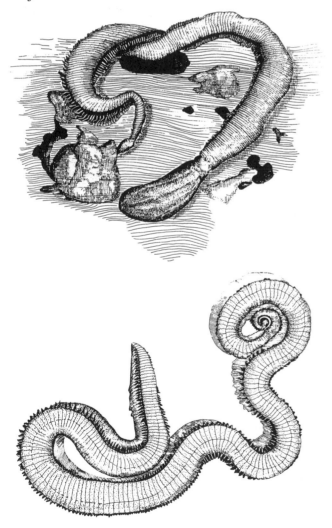

Figure 5-6. *The carnivorous polychaete,* Glycera robusta. *The upper illustrations shows the worm with its proboscis everted (extended).*

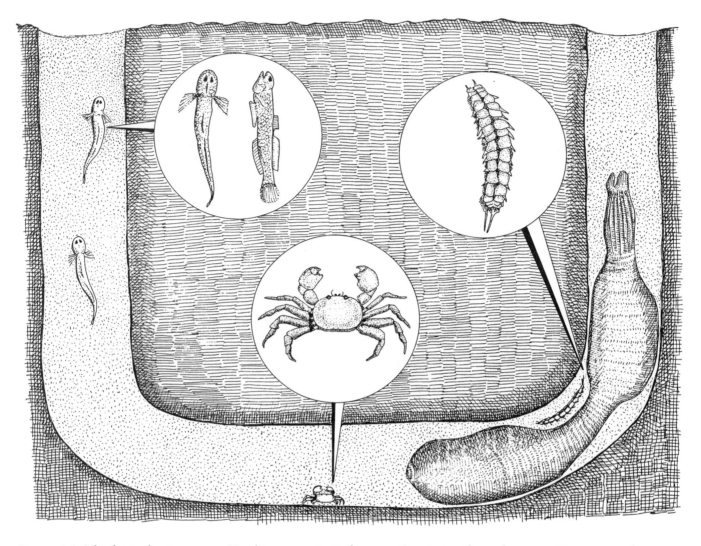

Figure 5-8. The fat innkeeper worm, Urechis caupo, *in its burrow. Guests are the scale worm,* Hesperone adventor, *the pea crab,* Scleroplax granulata, *and the arrow goby,* Clevelandia ios.

adventor (Figure 5-8) and a small (less than one half inch wide) pea crab, *Scleroplax granulata* (Figure 5-8), that also lives in the burrow. A third resident of the inn, a small (one inch long) fish known as the arrow goby, *Clevelandia ios*, only uses the burrow for protection during low tide.

When digging in sandy mud flats, you may come across long, narrow, flat worms that lack segmentation. These are probably nemertean worms of the Phylum Nemertea. Nemerteans are known as rubber band or ribbon worms, and their ability to elongate their bodies makes them successful burrowers. Nemerteans possess an eversible feeding structure and are very adept carnivores on other worms and

small crustaceans. Illustrated here is *Cerebratulus californiensis* (Figure 5-9), a relatively common worm in sandy mud. It is usually dirty-pink in color and reaches a length of three yards when extended! It is probably a good idea not to pick up any nemerteans you find because they fragment into many pieces when handled. As with all the worms discussed here, gently bury them in a shallow depression and they'll take care of the rest.

Mud Flat Clams. Clams are very much at home in protected, soft substrate habitats. The relatively undisturbed nature of the substrate allows them to sit vertically in semi-permanent burrows. Their posterior,

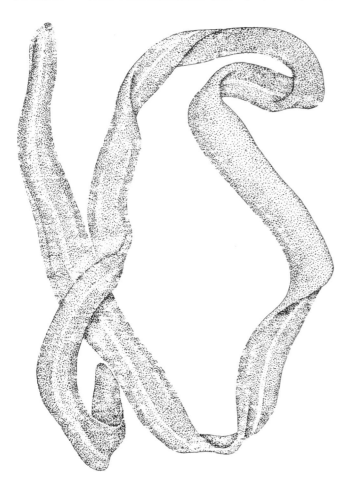

Figure 5-9. Cerebratulus californiensis, *a common large (up to three yards!) nemertean worm of California coastal flats.*

elongated siphons allow them to reach the water for suspension feeding while keeping their body safely beneath the sediment surface. The siphons are especially effective for clams that live in very muddy substrates where there is little water circulation below a very shallow surface sediment layer. Beneath this layer, oxygen is depleted by decomposing bacteria, and the substrate takes on a dark brown to black color that has the odor of rotten eggs. Obviously this isn't a hospitable place for an aerobic (oxygen consuming) organism, but it's perfect for clams that can burrow safely into this anaerobic mud and reach the surface with their siphons. Often the shells of clams taken from such a substrate will be stained black.

Sometimes when beachcombing a mud flat, you may be startled by a sudden spout of water squirting out of the substrate. Stomping your foot may produce several such spouts in the immediate vicinity. This is most likely the work of the large gaper or horseneck clam, *Tresus nuttallii* (Figure 5-10). The siphons of the gaper clam are so large that they can not be fully retracted into the shell, so the posterior end of the shell gapes open to allow the clam to close it shells around the rest of its body. This clam's shell can reach a length of over eight inches, and the clam can weigh upwards of four pounds. A clam this size can burrow over three feet into the sediment. The tell-tale sign that it is a gaper clam is the leathery patch that protects the siphon tips that sometimes still project above the substrate surface. With any disturbance, including the tread of hip waders, the clam quickly retracts the siphons into its burrow, and the water contained within them is forced out to produce the water spouts previously mentioned.

While the gaper is a real master burrower, it is not the champion of the mud flat. This honor goes to the geoduck, *Panope generosa*. The common name, geoduck, is derived from the native American name for this immense (up to a yard long including the siphons) clam that can weigh as much as 12 pounds. The siphons of *Panope* (Figure 5-11) are so large that the eight-inch long shells gape widely to accommodate them. Geoducks can be buried up to five feet in the substrate, and only the most determined, well-equipped beachcomber will unearth one. Geoducks are found in mud flats from British Columbia to central California, but are not abundant. In the Pacific Northwest they are taken commercially from subtidal clam beds using hydraulic digging gear.

Figure 5-10. The gaper clam, Tresus nuttallii. *The siphons can not be fully retracted into the shells.*

Figure 5-11. The geoduck clam, Panope generosa. *The shells gape widely around the large siphon.*

When excavating for the deeply buried gaper and geoduck clams, you may uncover another relatively large (up to five inches) mud flat calm. This is the Washington or butter clam, *Saxidomus nuttalli* (Figure 5-12). Washington clams are also deep burrowers and may have black-stained shells. The shells are ornamented with a continuous series of raised concentric rings and were used as money by native American tribes. The Washington clam occurs from Humboldt county to Baja California. A similar-looking, close relative, *Saxidomus giganteus*, occurs in the Pacific Northwest and reaches as far south as central California. Both species are known by the common names Washington or butter clam.

One of the more interesting mud flat clams is the bentnosed clam, *Macoma nasuta* (Figure 5-13). The bentnosed clam lies burrowed on its left side with the bent posterior end of the shell facing upwards. The yellow siphons of this clam are long, thin and separated from one another. The incurrent siphon, which brings water into the clam, probes about the substrate surface for deposited organic matter. The siphon vacuums the food into the mantle cavity where it is trapped on the surface of the gills in the same manner as a filter feeding clam. The bentnosed clam quickly vacuums up any available food on the surface and must move to a new spot. This is accomplished by a large, thin, digging foot that allows the bentnosed clam to burrow sideways in the sediment. Bentnosed clams are usually two inches in length (four-inch specimens are occasionally seen), and are found in bay and estuarine mud flats all along the coast from Alaska to Baja California.

Another common mud flat species is the softshell clam, *Mya arenaria* (Figure 5-14). This six-inch-long

Figure 5-12. The Washington or butter clam, Saxidomus nuttalli. *Note the strong concentric rings on the shell in the foreground.*

Figure 5-13. The bentnosed clam, Macoma nasuta, *two inches long. Note the upward bend in the shell of the upper specimen.*

species was probably introduced into West Coast estuaries along with the Virginia or blue point oyster, *Crassostrea virginica* (Figure 5-15). The oysters were initially brought around the horn of South America in clipper ships during the California gold rush to satisfy the pallet of newly rich 49ers. When the transcontinental railroad was completed, oysters were brought in by the train-car load along with a plethora of uninvited hitchhikers. The softshell was seen as a welcome addition by many, as the clam is able to live in low-salinity estuarine waters, grows rapidly, and is

delicious. It is now one of the most common bay clams from Alaska to Elkhorn Slough. The thin, brittle shells of the softshell clam are chalky white with a brown, flaky layer, called the periostracum, along the hinge region. When the shells disarticulate (come apart) as seen in the photo, the right-hand valve has a spoon-shaped shelf projecting outwards that strengthens the hinge region in life. It is the only common clam in California that has such a structure.

Often when the burrows of innkeeper worms, ghost shrimps, or large clams are excavated, a small (one half inch long), dull-white clam appears, sometimes in abundance. This is most likely the false mya, *Cryptomya californica* (Figure 5-16). *Cryptomya* is initially an enigma because it has very short siphons yet occurs very deep in the substrate. However, this clam taps into the burrows of other large invertebrates, and uses the water they circulate through their burrows for its own filter feeding. *Cryptomya* is common all along the California coast.

The California jackknife clam, *Tagelus californianus* (Figure 5-17), is sometimes locally abundant on southern California sandy mud flats. This four-inch-long clam has slender, straight, elongated valves that are thin and flat. It has a very active digging foot that allows it to move rapidly up and down its burrow and avoid harm. The jackknife clam occurs from Humbolt county to Panama, but is only common on sand flats from Santa Barbara south.

Figure 5-14. Three-inch-long shells of the softshell clam, Mya arenaria. *Note the conspicuous, spoon-shaped projection on the rear shell.*

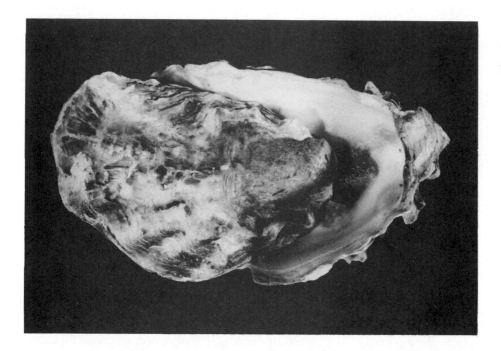

Figure 5-15. Four-inch-long shells of the Virginia or blue point oyster, Crassostrea virginica.

Figure 5-16. The false mya, Cryptomya californica, *a half-inch clam with short siphons found in the burrows of other large invertebrates.*

Mud Flat Snails. There are several obvious snails to be found on mud flats. The first of these is a southerner, Gould's bubble shell, *Bulla gouldiana* (Figure 5-18). As the name bubble implies, *Bulla* has a thin, inflated brown shell that can reach a length of two inches. The snail can not retract completely into its shell. *Bulla* is probably an omnivore, and can be found patrolling the surface of mud flats in the backwaters of coastal embayments like Mission Bay in San Diego County all the way north to Magu Lagoon in Los Angeles County.

The barrel-shell snail, *Rictaxis puntocaelatus* (Figure 5-19), occurs on the mud flats of central California bays, occasionally in large numbers. This deposit-feeding snail is less than an inch long, but is clearly marked with narrow black bands that wind around the shell.

The California horn snail, *Cerithidea californica* (Figure 5-20), is common from Tomales Bay in Sonoma County south to Baja California. It reaches a length of one and a half inches. Its tapering, buff to brown colored shell is unmistakable. The horn snail can be found on mud flats including those in estuaries. However, it is most common in salt marshes, where it aggregates in tremendous numbers in tidal creeks and high salt marsh pools called salt pans.

In San Francisco Bay the horn snail has been literally chased off the mud flats by the introduced mud snail, known as *Illyanassa obsoleta* or *Nassarius*

Figure 5-17. Shells of the jackknife clam, Tagelus californianus. *Shells are four inches long and one inch high.*

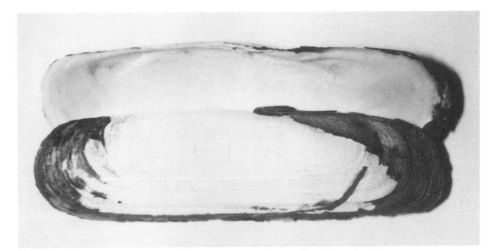

Figure 5-18. Gould's bubble shell, Bulla gouldiana. *Specimen on left is two inches long. In life the snail can not retreat fully into its thin shell.*

Figure 5-19. The barrel-shell snail, Rictaxis puntocaelatus, *less than an inch in height.*

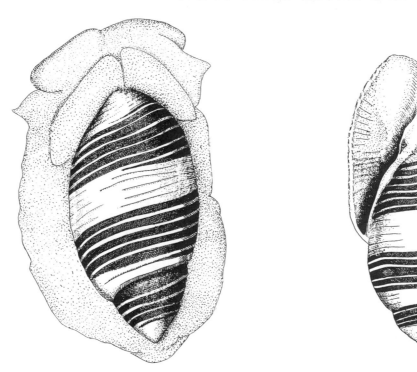

Figure 5-20. The tall, tapered shell of the California horn snail, Cerithidea californica.

obsoletus (Figure 5-21). The deposit-feeding mud snail was probably introduced along with the Virginia oyster (Figure 5-15), and has taken up residence with a vengeance. Mud flats and salt marsh channels throughout the estuary are peppered with these one-inch-long snails with the dull black to brown shells. The female mud snail packages her fertilized eggs into eighth-inch-high, vase-shaped egg cases, which she attaches to any hard surface. It was probably such egg cases, attached to oyster shells, that led to the mud snail's introduction.

A close relative of the mud snail is the native snail species known as the channeled basket shell, *Nassar-*

ius fossatus (Figure 5-22). This beautiful snail, which reaches two inches in shell length, is found in sandy mud habitats all along the West Coast. In southern and central California it is often found on intertidal flats, at mid-tide or lower. In the spring, look for the three-eighths-inch-long egg cases laid by female snails attached to any handy solid substrate like a bottle, empty shell, or a sand dollar skeleton. *Nassarius fossatus* is a carnivore/scavenger and a very lively, fast-moving snail. If you have a plastic bag, immerse this snail in sea water and watch as it quickly comes out of its shell and begins to explore its surroundings. Look for the long, agile siphon that the snail uses for breathing and testing the water ahead for the presence of prey.

A delightful snail to look for all along the California coast is the long-horned nudibranch, *Hermissenda crassicornis* (Color Plate 10b). *Hermissenda* is one of the most common of the shell-less sea slugs. It is found in the rocky intertidal zone and on pier pilings and dock floats. *Hermissenda* occasionally shows up on sand and mud flats, usually in tandem with large amounts of drifting green algae that wash up and become stranded on the intertidal flats. This nudibranch eats a wide variety of prey including carrion (dead animal tissue) and can grow to three inches in length. It is readily distinguished by an orange area flanked by vivid light-blue lines running down the center of its back. The edge of the foot is likewise highlighted with light-blue lines.

Figure 5-21. The mud snail, known as Illyanassa obsoleta *or* Nassarius obsoletus, *an import from the East Coast. Note the elongated siphon.*

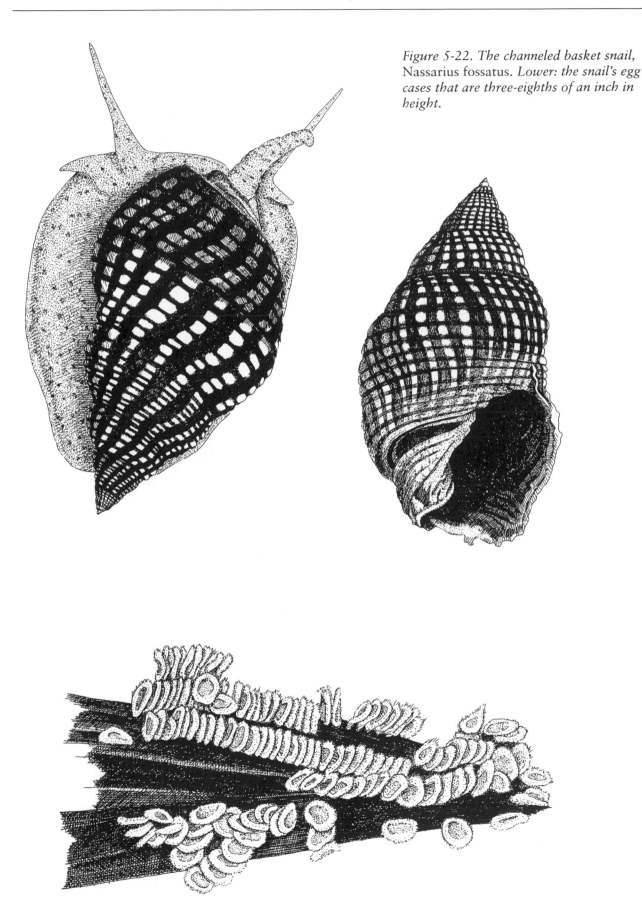

Figure 5-22. The channeled basket snail, Nassarius fossatus. *Lower: the snail's egg cases that are three-eighths of an inch in height.*

Another shell-less snail of low intertidal mud flats and eelgrass beds is *Navanax inermis* (Color Plate 9d). *Navanax* is a voracious predator, swallowing other snails like Bulla and *Hermissenda* whole. *Navanax* varies in color with a dark, velvet-brown base color and white or yellow spots or streaks. The margins of the foot and rolled mantle are usually lined with yellow-orange, and highlighted with bright blue. This snail can reach up to eight inches long and is found from Monterey Bay south to the Gulf of California. In southern California it is also found in the low rocky intertidal zone.

Mud Flat Crustaceans. In southern California the boundary between the salt marsh and the mud flat is the home of the small (one and a half inches across the carapace), active fiddler crab *Uca crenulata* (Figure 5-23). Fiddler crabs are so named for the outsized claw of the male crab. The claw is waved in a unique pattern to attract females into his burrow during breeding season. Fiddlers roam the marsh channels and mud flats at low tide during dawn and dusk in search of particulate organic detritus on the mud surface. During high tide, they return to their burrows and hide from their predators. Once quite abundant, fiddlers are now seen occasionally on southern California mud flats like those of Newport Bay in Orange County. They do not occur any farther north than Playa del Rey in Los Angeles County.

A much more abundant and widely distributed mud flat crab is the mud crab, *Hemigrapsus oregonensis* (Figure 5-24). *Hemigrapsus* males reach up to two inches across the carapace, and their claws are proportionately much larger than those of the smaller females. This crab can be found under any debris on the mud flat or in shallow burrows during low tide, and roaming the mud flat during high tide. Like the fiddler crab, *Hemigrapsus* is also a detritus feeder, but will also take carrion when it is available.

In San Francisco Bay a new crab has made a recent appearance. This is the green crab, *Carcinus maenas* (Figure 5-25). The green crab was discovered in south San Francisco Bay in 1991, and has since been found in several other Bay locations. The green crab is native to Europe and was introduced to the East Coast long ago. How it arrived in San Francisco and where it came from is currently a mystery. The green crab reaches about three inches across the carapace and is a voracious predator of clams and other soft substrate invertebrates.

Figure 5-23. Fiddler crab, Uca crenulata. *This male's large claw is used in courtship rituals.*

Figure 5-24. The mud crab, Hemigrapsus oregonensis, *common on soft substrates of bays and estuaries.*

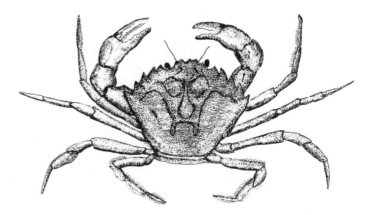

Figure 5-25. The green crab, Carcinus maenas, *introduced into San Francisco Bay from the Atlantic in the late 1980's.*

A fourth mud flat crustacean occurs beneath the sediment in burrows. This is the red ghost shrimp *Callianassa californiensis* (Figure 5-26). Ghost shrimps are very active burrowers and their excavations can reach two or more feet beneath the mud flat surface. As the shrimp burrows through the sediment it sifts and eats organic particles that were buried within the sediment when the mud flat was formed. The male shrimp is larger (up to three inches long) than the female and has one claw much larger than the other (Color Plate 13b). The shrimp uses his claw like a bulldozer blade to push sediment out of the burrow onto the surface. In the process he buries organic material that has been deposited on the substrate. The shrimps' continuous excavations can result in the upper layer of a local mud flat area being turned over several times a year.

The blue mud shrimp, *Upogebia pugettensis* (Figure 5-27), is closely related to *Callianassa,* but occurs lower in the intertidal zone. The blue mud shrimp is actually bluish white, reaches about six inches long, and lacks the large claw seen in male *Callianassa*. It is also a burrower, but does not need to burrow in order to feed. Instead *Upogebia* excavates a burrow and pumps water through it with its flat abdominal appendages. As the water flows over the shrimp's body, it strains suspended material with its legs and collects and swallows the food with its head appendages. Like many large invertebrate burrow dwellers, the blue mud shrimp may have "guests" or

Figure 5-26. The red ghost shrimp, Callianassa californiensis. *This three-inch-long specimen is a female.*

commensals living with them including pea crabs, scale worms, and the small arrow goby (Figure 5-8).

In addition to these large, obvious crabs and shrimp, many smaller crustaceans also occur on the mud flats. These include legions of small, tube-dwelling amphipods and burrowing isopods. These animals, along with numerous small worms, are the main food sources for the foraging shorebirds at low tide, and the small fishes and shrimp that come in at high tide.

Figure 5-27. The blue mud shrimp, Upogebia pugettensis, *a burrow-dwelling filter feeder.*

SAND FLATS

The Sand Flat Environment

As previously mentioned, sand flats are part of a continuum of soft substrate habitats. Many mud flat animals overlap into sand flats, especially the burrowers. Sand flats occur in protected environments where there is an appreciable flow of water and a source of sediment. For example in an estuary, sand flats typically form near the estuarine mouth adjacent to the main channel where relatively constant tidal currents keep the finer sediment particles in suspension and only the larger sand grains settle out. Sand flats are also common in coastal embayments (Figure 5-28) for similar reasons. The combination of tidal currents and locally generated waves transport sand into the embayment where it is dropped when the water movement slows. The most attractive feature of sand flats to beachcombers is they are firm and easily traversed. With a little patience, a good shovel, and a decent low tide, sand flats are very rewarding places to explore.

Sand Flat Cnidarians. On southern California sand flats the sea pansy *Renilla kollikeri* (Figure 4-26), introduced in the sandy beach chapter sometimes occurs by the hundreds. Look for them on sand flats as far north as Santa Barbara.

The moonglow anemone, *Anthopleura artemisia* (Color Plate 4b), can be found on sand flats from Alaska to southern California. However, this animal will always be anchored to a rock or some other solid substrate below the sand's surface. The moonglow anemone can reach a size of three inches across the tentacles and stretch upwards through the sand for over ten inches. Its color varies considerably, but the tentacles are usually transparent and brightly marked with either black, red, orange, white, or blue. It also occurs in the middle and low rocky intertidal zones.

Sand Flat Polychaetes. As previously mentioned in the mud flat section, many polychaete species overlap both sandy mud and sand flats. Lugworms, bloodworms, scaleworms, and most others can show up on sand flats.

The ice cream cone worm, *Pectinaria californiensis* (Figure 5-29), is a sand bottom specialist. *Pectinaria* uses the sand to construct a beautiful tube of individually selected sand grains. The cone-shaped sand tube fits snugly around *Pectinaria*'s tapered body. Sealing off the end of the tube are elegant, gold-colored bristles that the worm uses for digging. *Pectinaria* feeds using anteriorly positioned feeding tentacles that bring sand and organic detritus to the mouth. The tube may reach three inches in length and harbor a two-and-a-half-inch-long worm.

Figure 5-28. Sea gulls forage on a coastal sand flat at dusk. Princeton Harbor, San Mateo County.

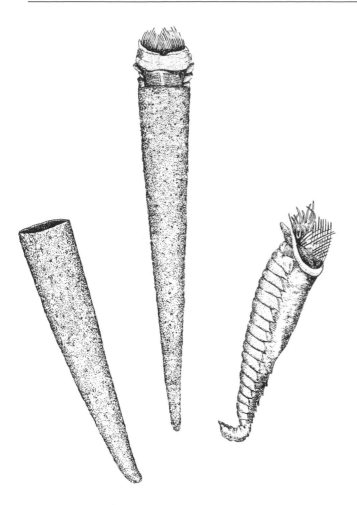

Figure 5-29. The ice cream cone worm, Pectinaria
californiensis. *Note the tapered body of the exposed
two-and-a-half-inch-long worm.*

A final sand flat polychaete sometimes seen is
related to the shimmy worm of the sandy beach,
Nepthys californiensis (Figure 4-10). This is *Nepthys
caecoides,* and it is a somewhat more robust version
of its beach-dwelling cousin. This three-to-four-inch-
long worm is a carnivore/scavenger and freely burrows through the sediment looking for food.

Other Sand Flat Worms. Fat innkeeper worms
(Figure 5-8) and ribbon worms (Figure 5-9) may also
occur on sand flats.

On some sand flats in central California, such as
those in Elkhorn Slough in Monterey County, or
Drakes Estero on Point Reyes in Marin County,
patches of substrate in the low intertidal zone seem
suddenly firm under foot. Careful excavation will
reveal the closely packed, flexible chitinous tubes of

the phoronid worm *Phoronopsis viridis* (Figure 5-30,
and Color Plate 6b). The phoronid bed may consist of
hundreds or thousands of these slender, three-to-five-
inch-long worms. When the tide covers them, a green,
double-spiraled filter-feeding structure, called a
lophophore, protrudes from the opening of the vertically oriented tubes. In southern California a similar-
looking phoronid, *Phoronopsis californica*, is found.
However, this worm does not occur in beds and has
an orange lophophore.

Figure 5-30. The phoronid worm, Phoronopsis virdis,
*in its sand-grain encrusted tube. The worm is three
inches long.*

Sand Flat Clams. The sand flat is a more physically active habitat than the mud flat. It is subjected to stronger water currents and sometimes, especially in coastal embayments, wave action. For the clams that live here this means that a burrow may occasionally be disrupted or buried. Most sand flat clams live shallower in the substrate than those on mud flats and have corresponding shorter siphons. To compensate for the occasional disruption that can sometimes wash them out of the sand, these clams have a very large digging foot and are superb burrowers.

The basket cockle, *Clinocardium nuttallii* (Figure 5-31), is an excellent example of a sand flat clam and can be found all along the California coast. This bivalve mollusk has a robust, heavily ribbed shell that can reach four inches in diameter. *Clinocardium*'s siphons are very short, barely protruding beyond the posterior end of the gaped valves. The foot is immense, stretching out over twice the length of the shell when fully extended. The cockle lives at such a shallow depth that it is easily excavated by the predatory sea star, *Pisaster brevispinus* (see Figure 5-40, discussed later). However, *Clinocardium* has a remarkable escape response up its sleeve (shell?). It protrudes its large foot and folds it under its shell, than rapidly stretches the foot its full length. The clam is pole-vaulted free of the substrate surface and out of the grasp of the sea star. A few more of these jumps and the clam has successfully avoided the slow-moving predator.

Sometimes co-occurring with the basket cockle is the common littleneck clam, *Protothaca staminea* (Figure 5-32). The white to tan, three-inch shell of this clam is also ribbed, but the ribbing is not as strong as *Clinocardium*'s, and is combined with a series of weak concentric rings. The common name "littleneck" is in reference to the clam's short siphons, and it is seldom buried deeper than three inches. *Protothaca* is not as good a digger as the basket cockle, so it is not found in habitats with shifting sands. Besides being found on protected sand flats, *Protothaca* also occurs in packed mud, and gravel mixed with sand and mud. In addition *Protothaca* will also settle into pockets of sand and gravel that collect in the rocky intertidal zone (Figure 5-33), sometimes becoming trapped in the rocks as it grows. *Protothaca*'s occurrence in the rocky intertidal zone has led to its other common name, the rock cockle.

A clam very similar to *Protothaca,* but from the other side of the Pacific, is the Japanese littleneck, *Tapes japonica* (Figure 5-34). *Tapes* was first found in San Francisco Bay in the 1930's and is now one of the most abundant clams in San Francisco and Tomales Bays. It is also quite common in Puget Sound, Washington, where it is harvested commercially. *Tapes* grows to about three inches in diameter and prefers a sandy mud to sandy sediment with some shell or rock mixed in. It is found in bays, sloughs, and estuaries from British Columbia to southern California.

Figure 5-31. The heavily ribbed shell of the basket cockle, Clinocardium nuttallii, *is unmistakable. This small specimen is one and a half inches across.*

Figure 5-32. A small (one-inch) littleneck clam, Protothaca staminea. *Note the short siphons that give it its common name, "littleneck."*

Figure 5-33. A large (two-and-a-half-inch) littleneck clam, Protothaca staminea, *found in a sand pocket in the middle rocky intertidal zone.*

Figure 5-35. Shells of the clam genus Chione. Clockwise from the largest shell (three inches in diameter): Chione undatella; C. undatella *with pigmented shells;* C. californiensis.

Figure 5-34. The Japanese littleneck clam, Tapes japonica. *This introduced clam is found in bays and estuaries all along the West Coast.*

Southern California sand flats from Ventura County to Baja California often harbor clams of the genus *Chione* (Figure 5-35). Like *Protothaca* and *Tapes, Chion*e's shells are ornamented with both radiating ribs and concentric rings. However, the rings and ribs are equally prominent, giving these three-inch-diameter shells a pronounced, basket-weave appearance. Known as hardshell cockles, there are three species of *Chione* found in protected soft bottom habitats of southern California, and they are all very similar in appearance. Little information is available about the biology of these clams, which are favorites of sport clammers. *Chione* species are also

Figure 5-36. Shells of the white sand clam, Macoma secta. *This clam is two and a half inches long.*

common on the bottom of tidal creeks of southern California coastal salt marshes.

The smooth, clean shell of the white sand clam, *Macoma secta* (Figure 5-36) is unmistakable. This clam is found in clean sand, up and down the California coast. Like its bentnosed cousin, *Macoma nasuta* (Figure 5-13), the white sand clam is a deposit feeder, and it has long, separated siphons and an agile digging foot. With this morphology, the sand clam can burrow down to a foot or more. It reaches up to four inches in length.

A final bivalve of southern sand flats is the thick or speckled scallop, *Argopecten aequisulcatus* (Figure 5-37). This animal has a three-inch-diameter shell that is either orange or reddish with dark blotches or spots. The shell has prominent ribs and "ears" on either side of the hinge. Although filter feeders like most bivalves, scallops can swim by rapidly opening and closing their shells. The thick scallop lies slightly buried or on the surface near the low tide level of sand flats. If disturbed while covered with water, it will quickly respond by swimming away.

Sand Flat Gastropods. The two most common snails on California sand flats are the olive snail, *Olivella biplicata* (Figure 4-11), and the moon snail, *Polinices* spp. (Figure 4-18). Both these snails were introduced in the sandy beach chapter. The olive sometimes can be found in large aggregations, either attracted to a common food source or in search of mates. Olives are most active at night when they can be found actively burrowing just beneath the sand, often with the top of their shell exposed.

Moon snails are solitary, usually showing up as a burrowed bulge in the sand or with just the top of their shell showing on the surface. When one of these large predatory snails is unearthed at low tide, it often has its huge foot extended. Gentle prodding of the foot will cause the snail to retract into its shell, but hold it well away from you as you may get wet! Unlike most mollusks which extend the foot internally by inflating special blood sinuses, the moon snails actively pump water into special chambers in their foot to help it expand. When the snail retracts into its shell, this water is forced out. The last thing visible as the snail retreats into the safety of its heavy shell is the horny operculum that closes off the entrance to shell and protects the snail's soft body.

Figure 5-37. Shells of the speckled scallop, Argopecten aequisulcatus. *Larger shell is two and a half inches across.*

Figure 5-38. Six-inch-diameter egg collar deposited by female moon smail, Polinices lewisii.

Figure 5-39. Shells of the channel whelk, Busycotypus canaliculatum, *an introduced species from the East Coast. Larger shell is three inches long.*

In the spring and summer beachcombers may find a sand-colored structure as big around as a saucer or small dinner plate that looks like a discarded plunger or plumber's friend. This is the egg collar of a moon snail (Figure 5-38). It is formed by the female who places her fertilized eggs within a matrix of mucus combined with sand grains. As the embryos develop within the collar, it gradually deteriorates so that it falls apart at the time the embryos are ready to swim away as planktonic larvae. As many as half a million embryos are contained in a single large egg collar.

If beachcombing on San Francisco Bay sand flats, you may see a large snail (up to seven inches long) with an elongated shell. This is most likely the channel whelk, *Busycotypus canaliculatum* (Figure 5-39). The channel whelk is another East Coaster that has made itself at home here, probably after hitchhiking in with the Virginia oyster. It is a clam predator that chips open the shell of its prey using the edge of its own shell as a chisel.

Sand Flat Crustaceans. First, some familiar animals that were introduced in the sandy beach chapter. The swimming crab, *Portunus xantusii* (Figure 4-20), of southern California and the graceful cancer crab, *Cancer gracilis* (Figure 4-21) from central California north, are both at home foraging on sand flats. They would be buried in the sand at low tide. Also from

central California north, juvenile Dungeness crabs (less than three inches in carapace width), *Cancer magister* (Figure 4-41), are often found on estuarine sand flats. This large, commercially important crab, which shows up on sandy beaches mainly as molted exoskeletons, will venture far up into estuaries in its juvenile stage, as juveniles are tolerant of much lower salinities than the adults.

Two other cancer crabs, *Cancer antennarius* and *Cancer productus* (Figures 7-8 and 7-9, respectively), are occasionally found on sand flats especially if there are some rocks or jetties nearby. However, these two large, brick-red crabs with black-tipped claws are much more common in the rocky intertidal habitat and are discussed in Chapter 7.

The combination of sand flats and nearby rocky areas will also yield the mud crab, *Hemigrapsus oregonensis* (Figure 5-24), and the larger (up to two and half inches in carapace width), green-lined shore crab, *Pachygrapsus crassipes* (Figure 5-49). You may find both of these crabs under the same rock lying on the sand flat surface. *Pachygrapsus* is a scavenger and also very common in the rocky intertidal zone and on pier pilings. These crabs can also be found together in the tidal creeks that drain salt marshes. *Hemigrapsus* excavates burrows in the creek banks, and *Pachygrapsus* usurps and enlarges them.

The salt and pepper shrimps, *Crangon* spp. (Figure 4-23), introduced in the sandy beach chapter, are also common on sand flats. They come in at high tide and forage for prey. If stranded, they burrow in and wait for the next high tide. The familiar mole crab of sandy beaches, *Emerita analoga* (Figure 4-4), is sometimes found on sand flats if any wave action occurs.

A large hermit crab (over an inch in carapace length) is found on the sand and mud flats from Bodega Bay in Sonoma County south to Baja California. This is *Isocheles pilosus* (Color Plate 13c), and it typically lives in *Olivella* and other snail shells when young, and *Polinices* shells when larger. It usually has the left claw larger than the right. Like all hermit crabs, it is a scavenger and forages on the intertidal flats for bits of organic detritus that it picks up with its claws. It burrows into the sand at low tide.

Sand Flat Echinoderms. As their name suggests, you would expect to find sand dollars on sand flats. The common Pacific sand dollar, *Dendraster excentricus* (Figure 4-24), is exposed on sand flats in southern California in the Newport-San Diego area, usually near a tidal channel or some other source of moving water. Elsewhere in California, sand dollars are much more common offshore of sandy beaches, as discussed in Chapter 4. When exposed on the sand flat, the sand dollars usually lie flat and are completely or partially burrowed.

A much more obvious echinoderm on sand flats is the sea star known as the pink or short spined pisaster, *Pisaster brevispinus* (Figure 5-40). This sea star is found in many habitats from sand flats and the rocky intertidal zone to pier pilings, and well offshore on subtidal sand and mud bottoms. It is known to occur from Alaska to San Diego. On sand flats *Pisaster* feeds primarily on bivalve mollusks. It has the ability to find and extract a burrowed clam from the substrate either by digging after it or using special tube feet located around its mouth. These tube feet can penetrate the substrate to a depth equal approximately to the length of the sea star's arm. The tube feet are equipped with suckers that attach to the clam's shell and allow it to be pulled up to the sea star's mouth. The pink sea star can reach a diameter of over two feet in subtidal habitats, but the sand flat individuals are usually less than a foot across.

There are other echinoderms sometimes seen on or in the sand flat. Worm-like sea cucumbers work through the sediment, swallowing it as they go much like earthworms.

Figure 5-40. The pink or short-spined pisaster, Pisaster brevispinus, *a common predator found on soft bottoms.*

SALT MARSHES

Protected coastal environments are typically created by the flow of sediment into a bay, river mouth, or other shallow area, forming a delta. As the delta builds up, it can be colonized by special plants adapted to sea-water inundated soils and a salt marsh is born. Once the salt marsh plants take hold, their spreading root and underground stem systems trap more sediment. The level of the marsh is raised and the margin of the marsh increases outward until it reaches a dynamic equilibrium with the local pattern of water movement that controls sedimentation. As the marsh becomes more extensive, drainage channels, called tidal creeks, develop that direct the flow of water out of the marsh as the tide recedes. These tidal creeks can become deeply eroded, and their bottoms and banks create a habitat for invertebrates.

The West Coast does not contain the vast acreage of salt marshes as is seen along the Southeastern Seaboard or the Gulf of Mexico. In addition much of the salt marsh habitat has been diked and drained and turned into pasture or commercial real estate in California. The value of salt marshes and other coastal wetlands has only recently become apparent to decision makers, and there is some move to preserve and even restore wetlands, including coastal salt marshes. What salt marsh habitat we do have is generally protected and we hope will be preserved.

The average beachcomber is not going to venture too far into a salt marsh. The vegetation is thick and the underfooting often unsure. Likewise the bottom of tidal creeks can be very muddy and hard to traverse. Many salt marshes in California do provide viewing platforms and some, like the Palo Alto Marsh on San Francisco Bay, have catwalks that go out over the marsh or trails built along the marsh uplands. The portion of the marsh that is affected by the tides is dominated by two plants. Along the lower intertidal a fringe of the tall (up to three feet) cord grass, *Spartina* spp. (Figure 5-41) can be found. This fringe may be fairly extensive, but eventually gives way to species of the shorter pickleweed or glass wart, *Salicornia* spp. (Figure 5-42), that grow at higher tidal elevations and typically dominate the salt marsh. The marsh is laced with meandering tidal creeks that usually drain out across a low intertidal flat. Above the pickleweed is a community of upland marsh plants, and I refer you to more detailed accounts for reference [7, 8].

Figure 5-41. A small stand of cord grass, Spartina foliosa, *abuts a mud flat. This tall grass is found along the outer edges of California salt marshes.*

The marsh itself is home to a mixture of terrestrial and marine organisms. The terrestrial component includes many insects, small mammals, and of course the birds. Among the plants are many small marine invertebrates including crustaceans and mollusks [6, 7]. The tidal creeks are invaded by many of the same animals that were previously discussed as occurring on mud and sandy mud flats.

EELGRASS BEDS

A final protected, soft substrate habitat that may be seen by the beachcomber is the eelgrass bed. Eelgrass, *Zostera marina* (Color Plate 3c), grows along the lower intertidal edge of sand and mud flats and in shallow, subtidal protected habitats. If the water is clear, the subtidal eelgrass beds can be quite extensive like those of Elkhorn Slough and Humboldt Bay. Eelgrass beds create important habitat for fishes, shrimps, juvenile crabs and water fowl.

The intertidal eelgrass beds accessible to the beachcomber will yield many familiar animals already described, including many of the clam and worm species. The matted root mass of the grass is a particularly good place to observe polychaetes. There are also a variety of attached animals that can be found here as well [6].

I wish to introduce you to only one animal, and only an occasional visitor at that. This is the Califor

Figure 5-42. The low-growing *pickleweed or glass wart,* Salicornia *sp., the most common plant in California salt marshes.*

nia sea hare, *Aplysia californica* (Color Plate 9c). The sea hare is a large (up to 20 inches long), herbivore that can be found on a variety of subtidal bottoms. It eats a special variety of red algae and stores the noxious plant chemicals in its own body rendering it very distasteful to predators. Sea hares come into eelgrass beds to mate, sometimes in large numbers. They are simultaneous hermaphrodites, meaning they have both male and female reproductive systems at the same time. After exchanging sperm, the sea hares extrude long, sticky strings of jelly-encased eggs that they attach to the eelgrass. With their egg masses securely anchored the sea hares return to the subtidal zone to resume their feeding. After some weeks of developing the embryos escape from the jelly egg string and swim away as planktonic larvae.

Several other unique as well as common invertebrates occur in the eelgrass beds. However, the average beachcomber is not going to visit them very often as they are out of the way, and usually very low in the intertidal zone. For further information see References 8 and 9.

ROCKY SHORES OF PROTECTED COASTAL HABITATS

This chapter is intended to introduce habitats with soft substrates. However, in many of these environments some amount of rocky substrate occurs, either natural or man-made. Usually the organisms found here are a subset of the open coast rocky intertidal zone, and that chapter should be consulted. However, there are a few unique animals that occur on these protected rocky shores and these are mentioned briefly here.

In California estuaries, a small, native oyster is found. This is the Olympia oyster, *Ostrea lurida* (Figure 5-43). This bivalve reaches about two inches in diameter and can be found on the undersides of rocks, on concrete pilings, as well as attached to the discarded shells of other mollusk species. Oyster larvae prefer a shell substrate for settlement and sometimes this oyster will be found in small aggregations, but nothing like the huge shell reefs formed by some oyster species along the East Coast.

Often a snail is found close by the native oyster with a pretty, ornamented, inch and a half-long shell. This is the Atlantic oyster drill, *Urosalpinx cinerea*

Figure 5-43. Top and bottom shells of the Olympia oyster, Ostrea lurida. *Top shell is one and a half inches long.*

(Figure 5-44), which was introduced unintentionally with the Virginia oyster. This snail is now found in estuaries from British Columbia to Newport Bay. It feeds chiefly on barnacles and bivalves, and occasionally on other snails. As the common name implies, this snail uses its radula to drill a hole in its prey in the same manner as the moon snail.

A common mollusk on these protected rocky shores is the blue or edible mussel, *Mytilus edulis* (Figure 5-45). This wedge-shaped bivalve is found around the world in temperate waters. It is the common rocky intertidal mussel on the East Coast. Along the West Coast, it is typically much more abundant in protected bays and estuaries. The edible mussel is usually about two inches long although four-inch individuals are found. It does occur occasionally on

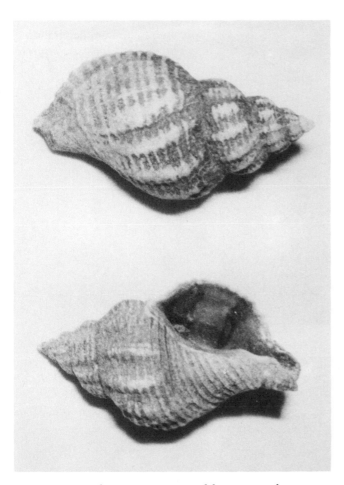

Figure 5-44. The East Coast snail known as the oyster drill, Urosalpinx cinerea. *The shells are one inch in height.*

Figure 5-45. The blue or edible mussel, Mytilus edulis, *two and a half inches long. Note the "beard" of attachment fibers, called byssal threads.*

the open coast usually with the much more robust California mussel, *Mytilus californianus* (Figure 6-56). Mussels attach to the substrate using special hairs made of protein called byssal threads. These are attached by the foot directly to the rock or, in mussel clumps, to one another, and anchor the mussel solidly. When large clumps of mussels accumulate, they become safe havens for scores of small crustaceans, worms, and other small invertebrates.

Another mussel found exclusively in estuaries is the ribbed mussel, *Ischadium demissum* (Figure 5-46). The ribbed mussel is another East Coast introduction and is found occasionally attached to rocks along with *Mytilus edulis*. However, this four-to-five-inch-long bivalve is much more common attached among the roots and stems of salt marsh plants. In addition to the prominent ribbing, worn areas of the shell often have silvery highlights. The ribbed mussel

is known to occur in San Francisco Bay, Los Angeles Harbor, and Newport Bay.

On rocky substrate in the quiet backwaters of Newport Bay large aggregations of an interesting snail occur. This is the tube snail, *Serpulorbis* sp. (Figure 5-47). The shell of the tube snail is elongated and cemented to the rock. In large aggregations the shells become intertwined and a dense maze is created, which small creatures inhabit. The shells shown in Figure 5-47 are cleaned museum specimens; in life the shells would be teeming with marine growth. Tube snails are filter feeders. They spread triangular-shaped mucous nets with their extended feet to trap suspended food from the water column. Tube snails are common animals in the southern California open rocky intertidal habitat as well, where they often carpet the undersides of large rocks and line the walls of wave-cut surge channels.

Figure 5-46. The ribbed mussel, Ischadium demissum, an introduced species common in salt marshes.

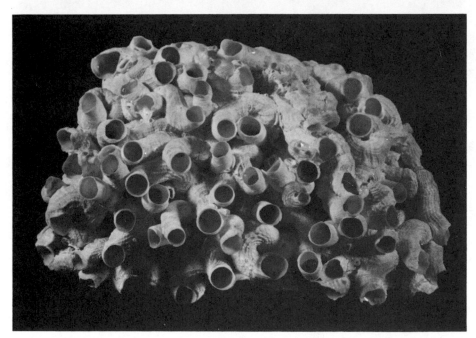

Figure 5-47. A five-by-six-inch mass of the intertwined shells of the tube snail, Serpulorbis squamigerus, a worm-like gastropod.

Another common rocky intertidal mollusk found in association with the tube snail is the rock oyster *Chama arcana* (Figure 7-23). This two-inch-diameter bivalve attaches solidly to the substrate with its left valve. The right valve is ornamented with wavy frills that are often eroded or fouled with marine growth.

A final animal must be mentioned, just because it is such a celebrity. This is the fish known as the plain-fin midshipman, *Porichthys notatus* (Figure 5-48).

This 15-inch-long fish has a double row of white, light-producing organs (photophores) along its underside that look like the buttons on the front of a junior marine officer's trousers, and thus the common name "midshipman." *Porichthys* is also known as the singing toadfish because of the loud croaking sounds it produces during mating season. The croaking is the reason for the midshipman's notoriety. The fish comes into estuaries to mate in the late spring. The

Figure 5-48. Side and bottom views of the singing toadfish or midshipman, Porichthys notatus, *that can reach 15 inches in length.*

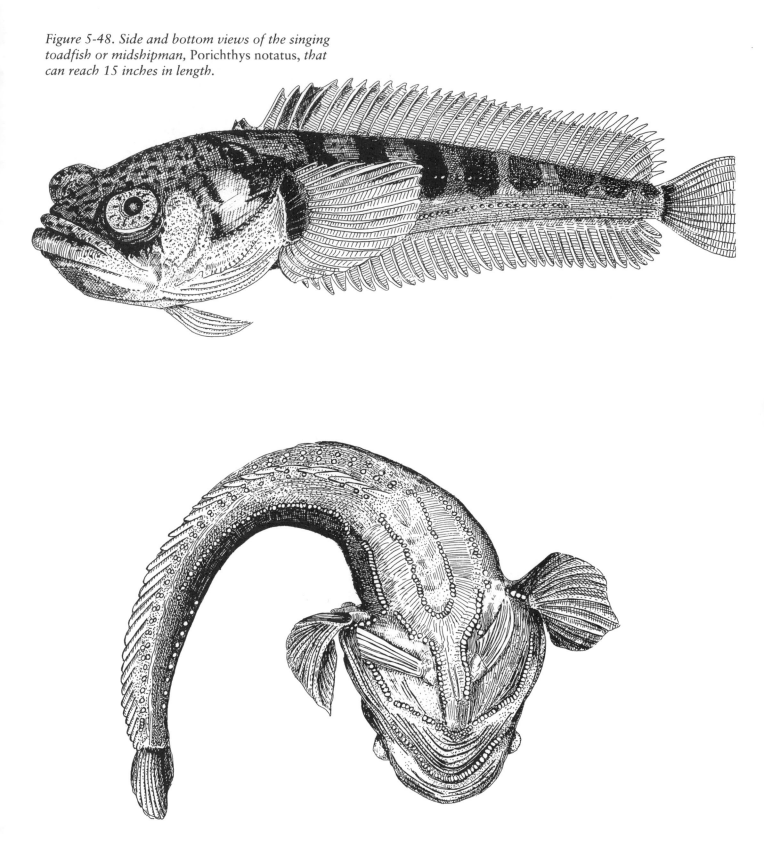

males set up territories near rock piles and, when
night falls, proceed to croak their presence to the
females. A female comes in and lays her pea-sized
eggs in a tight cluster under a rock that the male will
then guard until they hatch. These rocky nursery areas
are sometimes in the low intertidal zone, and an
investigating beachcomber may discover a guarding
male and his brood of eggs under a rock. After a
quick look carefully replace the rock and let daddy do
his duty.

Now for the interesting part. The singing toadfish
produces its croaking song by rapidly contracting the
muscles attached to its swim bladder, an internal gas
chamber that fish use to adjust their buoyancy. The
muscle contractions set up a rhythmic vibration in the
swim bladder, which acts as a resonating chamber to
produce a deep, almost mechanical hum. A few years
ago a group of love-starved toadfish set up camp and
started singing beneath the houseboats in Sausalito,
on San Francisco Bay; they created quite a stir. The
combined humming of the toadfish was heard at night
through the keels of the houseboats. Experts on
underwater sound were called in. Speculation on the
source of the alien sound ranged from miniature
Russian submarines on clandestine, nocturnal forays
to illegal pumping of pollutants into the bay. The
mystery sound made the local newspapers and,
because it was one of the slow-news days of summer,
it was picked up by the wire services and made
national news. The reaction when I identified the cul-
prit fish gave me my 15 minutes of fame, and made
the toadfish a local hero. There is now an annual
toadfish festival, complete with parade, every spring
in Sausalito.

OTHER HARD SUBSTRATES

In addition to the unique marine and estuarine
organisms discussed here, quiet-water habitats attract
human animals as well. Estuaries and coastal embayments are the preferred location for harbors, marinas,
fish-processing plants, etc. Consequently, you will
find a number of hard substrates provided by humans
in quiet-water habitats. Pier pilings and floating boat
docks are among the most common, and I will briefly
discuss them here.

Pier Pilings. These structures represent a hard substrate that receives normal tidal exposure. As a result,
they harbor a variety of organisms that prefer and/or

Figure 5-49. A large green-lined shore crab,
Pachygrapsus crassipes, *clings tenaciously to a
barnacle-encrusted pier piling (see also Figure 6-8).*

tolerate these conditions. The upper part of a piling is
usually the domain of hardy barnacles and a smatter-
ing of limpets typical of the upper intertidal zone of
the rocky coast (see Chapter 6). Also found here is
the green-lined shore crab, *Pachygrapsus crassipes*
(Figure 5-49). Below the barnacles, at approximately
middle tide level, the blue mussel, *Mytilus edulis*
(Figure 5-45) will be common, often in sizable
clumps. Mixed in with the blue mussel may be the
open coast mussel *Mytilus californianus* (Figure 6-
56). The mussels provide attachment sites for a variety of organisms including barnacles, limpets, sea
anemones, and encrusting forms like sponges and
bryozoans. If the clump is especially well developed,
it will harbor sea star predators and myriad small
invertebrates. See the treatment of the mussel clump
habitat in Chapter 6.

Floating Docks. The sides and bottoms of floating boat docks are the other common hard substrates you may find in quiet-water habitats. These are typically made up panels of high-density plastic foams that have high flotation properties and are imperious to sea water. However, they are not imperious to marine life! Unlike the pier piling, these substrates are never exposed to tidal action. Consequently, they harbor a suite of fragile, space-grabbing invertebrates, many of which grow in flat, encrusting colonies. Look for the bright reds and yellows of marine sponges. Also plentiful are the purple and orange colonies of compound tunicates, and the delicate, basket-weave patterns of encrusting bryozoan colonies (see Chapter 3). Erect or branching bryozoan colonies are also common along with a variety of hydroid cnidarians. Barnacles and mussels will also grow on these structures, and they provide yet more substrate for the encrusting species.

Several species of relatively large solitary tunicates do very well on these floating substrates. The trans-

Figure 5-50. A cluster of the tunicate known as the sea vase, Ciona intestinalis, *found commonly on submerged hard substrates in quiet water habitats.*

Figure 5-51. Several specimens of the delicate white anemone, Metridium senile. *The largest anemone is an inch and a half tall.*

parent tunicate known as the sea vase, *Ciona intestinalis* (Figure 5-50), occurs all along the California coast. Individuals three to five inches long can be very abundant in local situations. This animal can pump over five gallons of water a day through its body for filter-feeding and respiration. Also look for the leathery tunics of the Monterey tunicate, *Styela montereyensis* , and its close relative, *Styela clava* (Figure 7-54).

Small sea anemones can also be very common, especially the delicate, translucent-white *Metridium senile* (Figure 5-51). Also look for the brightly colored fans of feeding tentacles belonging to feather-duster worms (Color Plate 7d; Figure 7-57) protruding from their protective tubes.

References

1. Peterson, R.T. *A Field Guide to Western Birds,* 3rd ed. Houghton Mifflin, Boston, 1990.
2. Udvardy, M.F. The *Audubon Society Field Guide to North American Birds: Western Edition.* Knoph, New York, 1977.
3. Harrison, P. *Seabirds: An Identification Guide.* Houghton Mifflin, Boston, 1983.
4. Hayman, P. et al. *Shorebirds: An Identification Guide.* Houghton Mifflin, Boston, 1986.
5. Cogswell, H.L. *California Natural History Guides: 40. Water Birds of California.* University of California Press, Berkeley, 1977.
6. Ricketts, E. F. et al. *Between Pacific Tides,* 5th edition. Stanford University Press, Stanford, 1985.
7. Smith, R.I. and J. Carlton. *Light's Manual: Intertidal Invertebrates of the Central California Coast,* 3rd ed. Berkeley:University of California Press, 1975, 716 pp.
8. Zedler, J. B. *The Ecology of Southern California Coastal Salt Marshes: A Community Profile.* U.S. Fish Wildlife Service, FWS/OBS-81/54, 1984, 110 pp.
9. Joselyn, M. N. *The Ecology of San Francisco Bay Tidal Marshes: A Community Profile.* U.S. Fish Wildlife Service, FWS/OBS-83/23, 1983, 102pp.

6

The Rocky Intertidal Environment: Upper Zones

A beachcomber's first trip to California's rocky intertidal zone (Figure 6-1) is bound to be a memorable one. This is especially true if your beachcombing has been along the Gulf Coast, with its miles and miles of sandy beaches, or the Southeastern Seaboard, with its extensive marsh lands and estuaries. These soft-bottom habitats, though certainly impressive, do not prepare beachcombers for their first encounter with an extensive, rocky intertidal habitat. The diversity of the plant and animal life can be overwhelming. However, it can also be the beginning of an adventure that will last a lifetime, and still have you discovering new dwellers of this intertidal landscape.

PHYSICAL FACTORS AND HABITATS

The attempt to cover the California rocky intertidal habitat in a general guidebook has been a daunting task. As we'll see, the rocky intertidal habitat is really several habitats stacked vertically on one another along a gradient of tidal exposure. As reviewed in

Figure 6-1. A broad, rocky intertidal reef exposed at low tide. Seal Cove, San Mateo County.

Chapter 1, tidal exposure is one of the three main physical factors effecting the distribution of marine intertidal organisms. The other two are degree of exposure to wave action and the type of substrate.

Wave exposure is related to several factors. The geographical location of a rocky habitat will determine if it is directly in line to receive the full force of the Pacific's storm waves. Most of California's big storms come out of the west or northwest, with fewer coming from the south. The presence of offshore islands or submerged rocky reefs that come between the rocky intertidal habitat and the storm waves can blunt their force. Thus they can be important influences on the degree of local wave exposure. Finally, the impact of wave exposure is related to the type of rocks making up the rocky substrate.

Rocky intertidal substrates vary in their composition. The basic materials making up the rocky coastline you see today probably had their origin in another geological epoch under vastly different circumstances compared to their present situation. Some habitats are hewn from sturdy igneous rocks that originated from fiery volcanic activities. Other rocky habitats consist of sedimentary rocks, formed by pressure on the compacted layers of sands, silts, and clays that once formed the bottoms of coastal embayments or estuaries. Sedimentary rocks such as sandstones, siltstones, and mudstones are softer, and provide less secure attachment sites for barnacles, mussels, etc. These substrates are eroded by wave action in different patterns than are the igneous rocks. Sedimentary rocks are also more easily penetrated by boring organisms, which in turn weaken the rocky substrate and influence the way they will erode.

Very hard, dense, igneous rocks like basalts will resist erosion and tend to weather evenly. Sedimentary rocks are more easily broken up by wave action, and often intertidal boulder fields will be found, consisting of broken pieces of the reef surface. Likewise, the weathering of coastal cliffs can contribute loose rocky material to the intertidal zone at their base. In places where different types of rocks intergrade, an uneven pattern of erosion can result. Wave-cut surge channels, tunnels, and caves, shallow and deep tidal pools, upraised outcroppings, and fringing tidal reefs all can be formed from differential erosion patterns.

The extent of the rocky intertidal zone depends on the slope of the wave-cut bench. This slope is again related to the type of rock and the immediate geography of the area. Rocky intertidal areas are often found at the base of steep, rocky cliffs. These areas tend to be likewise steep with vary narrow, vertical intertidal habitats. Other rocky intertidal areas occur on broad, wave-cut terraces. These tend to have extensive rocky intertidal areas with a very gradual slope.

The result of the interaction of wave and substrate is a rocky intertidal habitat that can vary considerably from place to place. The reef may be simple bedrock with little diversity of habitat. In contrast it may consist of a mix of flat areas strewn with algae and boulder fields, upraised substrate, tidepools and surge channels, and contain myriad niches for organisms to inhabit.

ORGANIZATION OF ROCKY INTERTIDAL HABITATS

Intertidal Zonation Scheme

As previously mentioned, the presentation of the rocky intertidal habitat as a single entity would be an overwhelming task. In this guide I will describe the rocky intertidal habitat by zones as suggested by the pioneer marine biologist Ed Ricketts [1]. Ricketts describes the intertidal habitats found along the West Coast and provides a general scheme of intertidal zonation for rocky shorelines in which each zone is described separately. These zones and the tidal elevations relative to MLLW (Mean Lower Low Water, the zero point of the tide tables; refer to Chapter 2) that they typically encompass in California are the:

1. High intertidal zone, which includes the uppermost area wetted by the sea down to five feet above MLLW.
2. Upper intertidal zone, which includes the tidal elevations from five feet to two and half feet above MLLW.
3. Middle intertidal zone, which extends from two and half feet down to zero feet MLLW.
4. Low intertidal zone, which extends from zero feet MLLW down to the lowest level the tides reach.

This chapter describes the upper and high intertidal zones, and the tidepool, exposed-rock surface and

(text continued on page 105)

Plate 9a. The giant keyhole limpet, Megathura crenulata.

Plate 9b. Two top snails, Calliostoma annulatum *(right) and* C. canaliculatum.

Plate 9c. Small California sea hare, Aplysia californica.

Plate 9d. The predatory sea slug, Navanax inermis.

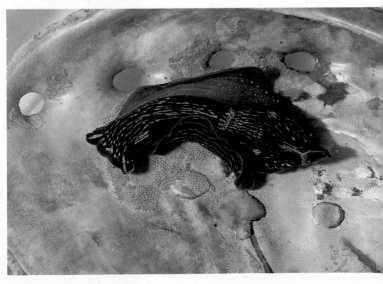

Plate 10a. The Spanish shawl nudibranch, Flabellinopsis iodinea.

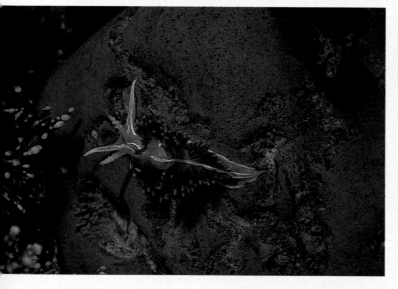

Plate 10b. The common long-horned nudibranch, Hermissenda crassicornis.

Plate 10c. The pugnacious nudibranch, Phidiana pugnax.

Plate 10d. The yellow-edged cadlina, Cadlina luteomarginata.

Plate 11a. Sea anemone predator, Aeolidia papillosa, typical color phase.

Plate 11b. An Aeolidia papillosa *that has been eating red sea anemones.*

Plate 11c.
Red sponge
with two
Leptasterias *sp.*
sea stars, and
two red sponge
nudibranchs,
Rostanga
pulchra.

Plate 11d. A red sponge nudibranch, Rostanga pulchra, *and its egg spiral.*

Plate 12a. The sea lemon nudibranch, Anisodoris nobilis.

Plate 12b.
The ring-spotted dorid,
Diaulula sandiegensis.

Plate 12c. The white spotted sea goddess,
Doriopsilla albopunctata.

Plate 12d. Triopha catalinae *(upper), and* Triopha maculata.

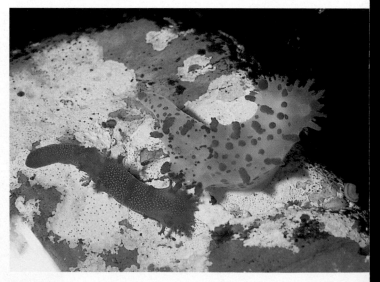

Plate 13a. Molted exoskeletons of the market or Dungeness crab, Cancer magister.

Plate 13b. Male (upper) and female red ghost shrimps, Callianassa californensis.

Plate 13c. The large soft bottom hermit crab, Isocheles pilosus. This specimen is living in the shell of a moon snail.

Plate 13d. The umbrella crab, Crytolithodes sitchensis.

Plate 14a. The sunflower sea star, Pycnopodia helianthoides.

Plate 14b. Two color forms of the Pacific sea star, Pisaster ochraceus.

Plate 14c. The giant pisaster, Pisaster giganteus.

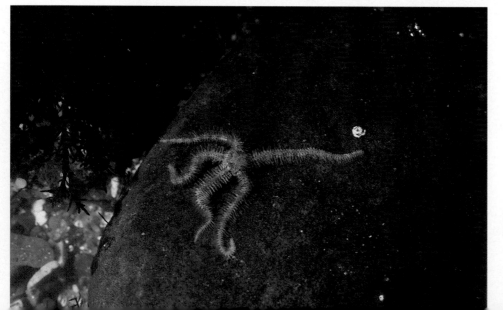

Plate 14d. The spiny brittle star, Ophiothrix spiculata.

Plate 15a. Several purple sea urchins,
Strongylocentrotus purpuratus.

Plate 15b.
The large red sea urchin,
Strongylocentrotus
franciscanus.

Plate 15c. The sea cucumber, Parastichopus
parvimensis.

Plate 15d. Small red sea cucumber, Lissothuria
nutriens.

Plate 16b. The Garibaldi, Hypsypops rubicundus, *the state marine fish of California.*

Plate 16a. A tidepool sculpin, Clinocottus *sp.*

Plate 16c. Two male elephant seals, Mirounga angustirostris, *square off.*

Plate 16d. Close quarters shared by juvenile California sea lions, Zalophus californianus.

(text continued from page 96)

mussel clump habitats of the upper and middle intertidal zones. The following chapter describes the remaining habitats of the middle and low intertidal zones, including those of the open reef flat, underrock, burrowing, low tidepool and surge channel.

As mentioned, exposure to wave action, tidal level, presence or absence of standing water, slope, rocky substrate type and erosion pattern, and many other factors contribute to the make-up of these zones. Therefore, although these descriptions are presented separately, you will observe that they are seldom discrete units, but instead intergrade and overlap one another. Similarly, many organisms will be present in several different zones, while others will be unique to only one.

Finally, there will be a group of organisms that are common all along the rocky coast of California and others that are restricted to some portion of it. I will present the most common organisms a beachcomber is likely to find all along the California coast. If any of the plants or animals I include is not found along the entire coast, I will mention where they are known to occur.

HIGH INTERTIDAL ZONE — UPPERMOST HORIZON

The highest rocky intertidal habitat recognized is called the uppermost horizon or Zone 1 by Ricketts [1]. This is the region covered only by the highest tides and the narrow strip above the high water mark that is still influenced by the sea, primarily by the splash from waves and wind-borne sea spray. This is a zone of transition between the land and the sea, and many of the organisms dwelling here take advantage of aspects of both environments (Figure 6-2).

Seaweeds. Prominent here are algae that depend on the sea spray and fresh water seeping out of cliffs that often back rocky intertidal zones. Bright green patches of the green alga, *Enteromorpha* sp. (Color Plate 1b), can be found in shady, moist areas. Another bright green alga seen here is the sea lettuce, *Ulva* spp. (Color Plate 1a), although it is much more common in the lower zones. *Ulva* spp. come and go quickly, often completing their life cycles in a few weeks. However, they can occur in great profusion throughout the intertidal zone, often growing on other algae as well as the rocky substrate.

A few hearty rockweeds hang on in the lower part of this zone. Algae of the high intertidal zone are more prominent in the winter when warm sunny days are few, and when sea spray, wave splash, and rain are plentiful, keeping this zone moist. In the spring and summer the high intertidal zone dries out, and the algae die back. Other algae grow in this zone, however they are small blue-green algae and diatoms that grow close to the rock surface and are not readily visible to us. These microscopic plants are most impor-

Figure 6-2. A vertical cliff face in the rocky intertidal zone. Note how the barren upper and high intertidal zones grade into the abundance of mussels and algae of the middle intertidal zone.

tant to the few animals that live here, primarily marine snails that feed on this thin algal film growing on the rocks.

High Intertidal Snails. The highest dweller in this zone is the periwinkle, *Littorina keenae* (Figure 6-3), which can be found in shady cracks in the rock. This snail can withstand exposure to air for up to three months as well as live totally submerged. Its shell can reach three quarters of an inch in height and is gray-brown in color. It has a white band on the interior of the aperture (opening) and the shell is often much eroded. *L. keenae* feeds on the fine film of small algae and diatoms that grow on high intertidal rocks. When exposed to air it can secrete a mucous holdfast around the aperture of its shell. This glues the animal to the rock surface and seals it off from excess drying. This snail falls prey to the carnivorous snail *Acanthina spirata* (see Figure 6-49 later in chapter) in the lower portion of its distribution where the two snails overlap.

Limpets also occur in this zone. Limpets are snails with broad conical shells that nestle flat against the rock to prevent water loss. On shady, vertical surfaces the digit or ribbed limpet, *Collisella digitalis* (Figure 6-4), can be found in small aggregations. This limpet is common on vertical rock faces in high and middle intertidal zones. It also occurs on mussels and stalked barnacles in mussel clumps in the middle intertidal

zone. The shell reaches a length of one and a quarter inches and the apex (top) of the shell is well forward, sometimes overhanging the shell margin (edge). *Collisella digitalis* receives the common name digit limpet because it looks like a flexed digit or finger when viewed from the side.

Another limpet seen in this zone is the rough limpet, *Collisella scabra* (Figure 6-5). This limpet is common in high and upper intertidal zones on horizontal and sloping rock faces. The shell is up to an inch and a quarter long, and the apex is forward of center; it is heavily ribbed with a scalloped margin.

Figure 6-4. A high intertidal limpet, Collisella digitalis, *one half inch long. Note how far forward the apex (top) of the shell protrudes.*

Figure 6-3. Five, half-inch-long specimens of the eroded periwinkle, Littorina keenae, *share the shade of a high intertidal crevice.*

Figure 6-5. *The high intertidal limpet,* Collisella scabra, *shown next to its excavated home scar. The limpet is three quarters of an inch long.*

The rough limpet exhibits "homing behavior." Using its shell and its rasping tongue (radula), the rough limpet excavates a depression in the rock, called a home scar, that exactly matches its scalloped shell. Like the digit limpets, these animals move about at high tide to feed on the microscopic film of algae on the rocks, but the rough limpets always return to their "home scar" to wait out the low tide period. Predators include shorebirds, the green-lined shore crab, *Pachygrapsus crassipes* (Figure 6-8), and sea stars.

High Intertidal Crustaceans. A pair of semi-terrestrial isopods, known as rock louses, treads the fine line between air and water in this zone. *Ligia occidentalis* (Figure 6-6), is the more obvious of the two

species. It spends most of the day hiding in crevices or under stones above the high tide mark. It emerges in the late afternoon to work the night shift. Like the limpets and periwinkles, this inch-long crustacean also feeds on the microscopic film of algae growing on the rocks. *L. occidentalis* is found from Sonoma County south to Baja California. Although it can not withstand continuous submersion, it must keep its breathing apparatus moist. The rock louse can be seen moving down to tidepools and dipping its rear end, which contains its gills, into the water. The second species, *Ligia pallasi* (Figure 6-7), occurs from Oregon to Santa Cruz County. Males of this species have greatly expanded side plates, and have always reminded me of mini-trilobites, those ancient relatives

Figure 6-6. *A rock louse,* Ligia occidentalis, *one and a half inches long. This large isopod moves rapidly over upper intertidal rocks.*

of the crustaceans. *L. pallasi* lives in crevices in cliffs above the high tide mark and is especially abundant in sea caves. It is slower and larger (one and a quarter inches long) than *L. occidentalis*, and is also a nocturnal feeder on algal films.

A more obvious crustacean, the green-lined shore crab, *Pachygrapsus crassipes* (Figure 6-8), is sometimes found here. This crab is found throughout the high and upper intertidal zones. It occurs in crevices in the high intertidal zone, under rocks and in crevices in the upper intertidal zone, and among the mussel clumps of the middle intertidal zone. Male *Pachygrapsus* reach a carapace width of two inches. The crab is dark in color and marked with shades of red, purple, or green. *Pachygrapsus* is an excellent scavenger, feeding on a variety of plant and animal material and occasional live prey such as limpets or *Littorina*. Its main food is the low-growing, microscopic algae that it scoops up with spoon-like claws. *Pachygrapsus* is a nimble-footed, aggressive species that challenges predators and humans alike with a pugnacious, claws-raised-and-outstretched stance. However, when given the chance, it quickly scuttles to the safety of a crevice or under a rock.

Small patches of acorn barnacles occur in the lower portion of the high intertidal zone. The small, volcano-shaped species, *Balanus glandula* (Figure 6-9), occurs throughout the intertidal zones on upraised substrates, and is at its physiological limit in the high

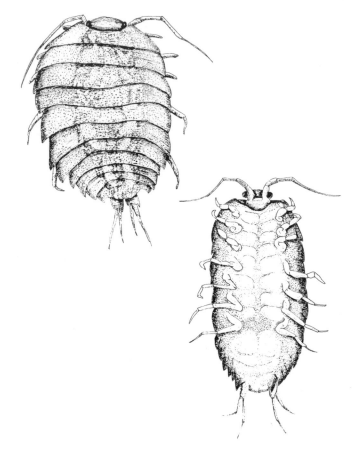

Figure 6-7. Top and bottom views of the rock louse, Ligia pallasi, *one and a half iunches in length.*

Figure 6-8. A small green-lined shore crab, Pachygrapsus crassipes, *nestled in an upper intertidal crevice.*

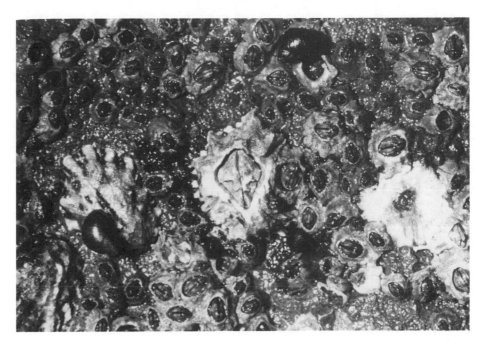

Figure 6-9. The common acorn barnacle, Balanus glandula *(center), surrounded by the smaller barnacle,* Chthamalus fissus, *a rough limpet, and periwinkles.*

intertidal zone. This species is the most common intertidal barnacle in California, and like all acorn barnacles, is a filter feeder. It occurs in large patches, and it is probable that its larvae are attracted to already-settled individuals, accounting for the dense settlements of these quarter-of-an-inch animals sometimes seen. Another acorn barnacle found here is the gray or brown *Chthamalus fissus* (Figure 6-9), which is about half the size of *Balanus glandula*. Acorn barnacles are eaten by sea stars and carnivorous snails.

UPPER INTERTIDAL ZONE

The upper intertidal zone (Figure 6-10) extends from five feet to two and half feet above MLLW. The prevailing physical factors operating here are exposure to air and sunlight during low tide. Thus the organisms that exist successfully in this zone are well adapted to withstand drying out (desiccation).

Upper Intertidal Seaweeds. Several hearty seaweeds are found here. The dark-red brillo-pad weed, *Endocladia muricata* (Color Plate 2d), grows in saucer-sized tufts and larger patches. *Endocladia* is a red alga and will dry to almost crispness at low tide; it quickly rehydrates when the tide returns. Another common red alga is the Turkish towel, *Gigartina* sp. (Figure 6-11; Color Plate 2d), so named for the nubby texture of its blades. In late spring and summer, small

Figure 6-10. The relatively barren rocks of the upper intertidal zone flank a high tidepool.

Figure 6-11. One of several species of the genus Gigartina, *red algae of the high and middle intertidal zones.*

Figure 6-12. A red alga known as the "salt sac," Hallosaccion glandiforme, *the largest "sac" is three inches long.*

salt-sac alga, *Hallosaccion glandiforme* (Figure 6-12), can be found here, sometimes in large patches. A little lower down, and especially on the sides of slightly raised surfaces, the brown algae known as rockweeds, *Fucus* sp. and *Pelvetia* sp. (Figure 6-13; Color Plate 3a), occur. These algae survive the exposure of the upper intertidal by secreting a slick muscilage coating that inhibits water loss.

Often occurring with the rockweeds in southern California is the green alga known as sponge weed or dead man's fingers, *Codium fragile* (Color Plate 1d). Sponge weed is the most massive of the green algae, reaching over a foot in length and clumps weighing several pounds. The plant has dark green to black finger-like fronds. A smaller green alga, *Cladophora* sp. (Color Plate 1c), grows here also in dense, pin-cushion sized, bright green tufts that hold water like a sponge. Look for them on flat surfaces often mixed among other algae.

Upper Intertidal Snails. Several marine snails occur in the upper intertidal zone all along the California coast. The small, black, checkered periwinkle, *Littorina scutulata* (Figure 6-14), occurs high in the upper zone, especially in depressions and cracks, and is common in both the upper and middle intertidal zones. It is very abundant in high tidepools and among mussel clumps. The shell is smooth, conical, and taller in proportion to diameter than *L. keenae.*

Figure 6-13. Fucus sp., one of several upper intertidal brown algae known collectively as "rockweeds."

Figure 6-14. Three small (three-eights of an inch high) checkered periwinkles, Littorina scutulata.

The shell reaches a half inch in height, although it is usually much smaller, and is brown to black, often with white markings in a checkerboard pattern. *L. scutulata* feeds on microscopic algae as well as the larger algae that co-occur with it including *Cladophora* (Color Pate 1c) and *Ulva* (Color Plate 1a). It is preyed upon by the snail *Acanthina spirata* (Figure 6-49) and the sea stars *Leptasterias* spp. (Figure 6-27).

A little lower in the upper tidal zone, the black turban snail, *Tegula funebralis* (Figure 6-15), makes its first appearance. The black turban is a herbivore, and is the most abundant and broadly distributed large snail in the rocky intertidal habitat. Smaller black turbans are found in the upper intertidal zone, with larger animals occurring lower in the intertidal. The black turban snail reaches a shell diameter of one and

a quarter inches. The shell is dark purple to black with silver white on the bottom. The spire (top) of the shell is often eroded, revealing the pearly shell layers beneath. The foot is black on the sides and pale beneath. *Tegula funebralis* eats several different types of algae including microscopic films growing on the rock surface, attached large algae, and drift plant material. *Tegula* is preyed upon by sea otters, rock crabs, and sea stars, especially *Pisaster ochraceus* (Color Plate 14b).

Tegula often has smaller snails living on the outside of its shell. The first of these is the small black limpet *Collisella asmi* (Figure 6-16). This limpet feeds on the tiny plants that grow on *Tegula's* shell. It changes *Tegula* shells when the turban snails aggregate during low tide. Studies have shown that each *C. asmi* changes *Tegula* hosts at least once a day. *Col-*

Figure 6-15. A small aggregation of the black turban snail, Tegula funebralis, *and two barnacle drills* (Acanthina spirata) *share a moisture-holding crevice.*

Figure 6-17. Two slipper limpets, Crepidula adunca, *on a* Tegula funebralis. *Note the charactertistic hooked shell apex.*

Figure 6-16. The small limpet, Collisella asmi, *that lives on* Tegula *spp. and other gastropods.*

Figure 6-18. Shells of the speckled turban snail, Tegula gallina, *of the southern California intertidal zone.*

Figure 6-19. The banded turban snail, Tegula eiseni. *Note the characteristic beading along the shell's bands.*

lisella asmi can grow to one half inch long, but is usually smaller. It is not unusual for a large *Tegula funebralis* to have more than one black limpet on board as well as one or more of a second snail, the hooked slipper limpet, *Crepidula adunca* (Figure 6-17), on its shell. This snail is also found occasionally on *Tegula brunnea* (Figure 6-25), *Searlesia dira* (Figure 6-26), and *Acanthina spirata* (Figure 6-49). The hooked slipper snail gets its name from the high, beak-like apex that overhangs the posterior margin of its shell.

This snail may reach a length of one inch, although it is usually smaller. Unlike the herbivorous *Collisella asmi*, *Crepidula adunca* is a filter feeder.

From Santa Barbara County south to Baja California, a second turban snail becomes common. This is the speckled turban snail, *Tegula gallina* (Figure 6-18). This snail looks very much like *Tegula funebrails*, and the two are very similar in diet and intertidal distribution. They are often found together where their ranges overlap. The speckled turban's

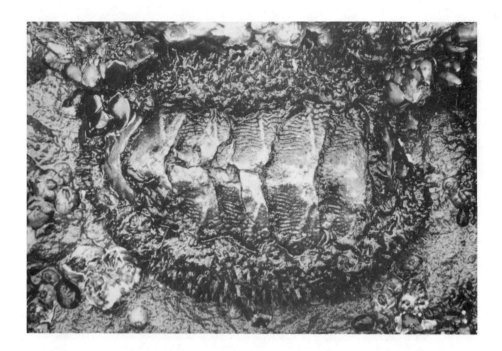

Figure 6-20. The mossy chiton, Mopalia muscosa. *Note how the shells are eroded.*

shell is pale gray to greenish and has stripes of white in a checkered or zig-zag pattern. This snail reaches a diameter of one and a half inches.

A third *Tegula* species is found from Los Angeles south. This is the banded turban snail, *Tegula eiseni* (Figure 6-19). This snail reaches up to an inch in diameter. Its shell is brownish and wound with beaded spiral ridges spotted with black. It is common in the middle and lower intertidal zones among rocks. One reference [2] reports that the banded turban is nocturnal, coming out at night to feed on algae.

The two limpet species of the high intertidal zone are also found here in the upper intertidal zone. The rough limpet, *Collisella scabra*, is seen occasionally on horizontal surfaces and the digit limpet, *Collisella digitalis*, is found on shady vertical surfaces.

Another mollusk common in the upper intertidal zone in California is the mossy chiton, *Mopalia muscosa* (Figure 6-20). Chitons are herbivores with their shells divided into eight plates. This species is usually found nestled up against a tuft of alga or some other organism that will harbor a reservoir of moisture during low tides. The mossy chiton reaches three and a half inches in length, and the shells are dull brown, although frequently eroded or covered with encrusting animals and plants. The tough, fleshy girdle that surrounds the shell plates is tan and is densely covered with stiff, round, reddish-brown bristles. The bristles give this chiton its "mossy" appearance. The

mossy chiton is active only at night during high tides when it feeds on algae such as *Endocladia, Gigartina,* and *Cladophora*. Many mossy chitons show homing behavior and move within a radius of about 20 inches from home base to which they return. Small mossy chitons in tidepools fall prey to sea stars, especially *Pisaster ochraceus* (Color Plate 14b).

Upper Intertidal Cnidarians. A widely distributed, intertidal cnidarian that makes its first appearance in the upper intertidal zone is the colonial, aggregating anemone, *Anthopleura elegantissima* (Figure 6-21). In most situations this anemone stops growing after it reaches an inch in diameter and increases its number by dividing itself in half, a process known as longitudinal fission. Because of this mode of reproduction, an aggregation of this anemone is a genetic clone, the result of repeated asexual divisions stemming back to the original pioneering anemone that grew from a planktonic larva. As a result, all the anemones in an aggregation have the same color pattern (Figure 6-21) and are the same sex.

High in the upper intertidal zone, *A. elegantissima* often occurs in a single-file line following the trace of a moisture-holding crevice. A little lower in the zone this anemone can be found in solid sheets on the shady sides of crevices and small upraised portions of rock, called outcroppings. When exposed at low tide, these colonial aggregations of anemones appear to be

Figure 6-21. Two identical clone mates of the aggregating anemone, Anthopleura elegantissima. *Each anemone is one inch across.*

Figure 6-22. A predator of sea anemones, this leather star, Dermasterias imbricata, *is five inches in diameter.*

layers of shell particles and other debris (Color Plate 4c). These materials are held by the retracted anemones to shield them from the sun and prevent water loss. When the tide returns, the anemones open up to form meadows of deadly, pink-tipped, flower-like polyps await for unwary prey.

Once you have identified clones of the aggregating anemone, look for places where a clear strip of rock separates two clones (Color Plate 4c). This space is a "no-man's land" that represents the area where two clones have come into contact, fought, and finally retreated. The anemones from the different clones actually attack one another, and fire their deadly stinging nematocysts in the battle. Eventually both clones will pull back and the clear strip remains as a buffer. I call it "clone wars."

Large, solitary individuals of *A. elegantissima* (Color Plate 4d), up to eight inches across the oral crown of tentacles, occur in the lower intertidal zones and in tidepools. These large animals are distinguished from other anemones by the conspicuous radiating lines on their oral disc. Anemones feed on zooplankton and small intertidal animals that are captured by their tentacles. They are preyed upon by shell-less snails known as nudibranchs, and the leather sea star, *Dermasterias imbricata* (Figure 6-22).

The upper intertidal zone will often include large numbers of acorn barnacles, *Balanus glandula* and *Chthamalus* spp. (Figure 6-9), especially on slightly vertical surfaces.

UPPER TIDEPOOLS

Tidepools are considered by many beachcombers to be the most interesting of all the intertidal habitats. Tidepools are formed by depressions in the rocky substrate that trap water at low tide (Figure 6-10). They thus provide a habitat that frees organisms from one of the most stressful factors of intertidal life, drying out. Therefore tidepools may contain a more diverse and often different association of organisms than found on adjacent, exposed substrate.

Physical Factors

However, tidepools are not without physical stresses, especially during low tide periods. A small volume of trapped sea water can experience a critical elevation in temperature on a warm, sunny day. Similarly, evaporation caused by wind and sun can raise the salinity to a sometimes dangerous level. Conversely, a very cold day can cause a decrease in water temperature during low tide, and a rainy low tide period may cause a reduction in salinity. The change in physical factors experienced within a tide pool habitat is related to the tidal level of the pool, its shape, and the volume of water it contains. A small, shallow pool located in the high intertidal zone would experience the greatest fluctuations, while a large, deep pool in the low intertidal zone would experience the least change.

Therefore, instead of visualizing tidepools as a single habitat, it becomes obvious that they constitute a continuum of habitats, depending on the particular situation of the individual pool. With this in mind, two tidepool habitats are described here. The first, called the upper tidepool, will encompass the tidepools of the upper, high and middle tidal zones, from approximately six to zero feet above MLLW. I will treat tidepools of the middle intertidal zone before I describe the organisms living on the other, exposed middle intertidal substrates as there is considerable overlap in the species found. A second tidepool habitat more characteristic of the lower intertidal zone is discussed in Chapter 7.

Tidepool Organisms

All the organisms described for the upper tidepool will not necessarily be found in every pool. The volume and tidal elevation, as well as the nature of the pool's bottom (i.e., gravel, cobble, sand, bare rock) will be important. However, the more common organisms are well represented. First of all, organisms overlapping from the upper intertidal zone will occur: *Tegula funebralis*, *Anthopleura elegantissima*, *Littorina scutulata*, and various limpets will be seen often.

Tidepool Plants. In all but the highest pools some form of coralline algae (Color Plates 2a and 2b) will appear. These red algae are called "corallines" because the calcium carbonate in their cell walls gives them a rigid, coral-like texture. Some coralline species occur as pale, pastel encrusting patches or sheets on the sides and bottoms of tidepools and on the rocks in tidepools. Other species grow upright from holdfasts and have joint-like regions of articulation along their length; these are called erect or articulated coralline algae. Because of their crusty texture, coralline algae are tough fodder for most intertidal herbivores and are not readily eaten. Therefore, they are more likely to be seen in tidepools compared to the softer, more palatable species of fleshy brown and red algae that can be more easily consumed.

Another plant that may be seen here is known as surfgrass, *Phyllospadix* spp. (Color Plate 3b). Surfgrass is not an alga. It is a flowering plant like the terrestrial grasses to which it is related. As such, it requires soil into which it can sink its roots to acquire the nutrients needed for growth. Therefore, you will find surfgrass only in pools with sediment on the bottom. Large patches of surfgrass occur in the middle and low intertidal zones as well.

How to Observe a Tidepool. The best way to investigate a tidepool is to approach it slowly and

Figure 6-23. The red-banded shrimp, Heptacarpus pictus, *one of several common small tidepool shrimps.*

avoid stirring the water. Observe the pool quietly for a few minutes and soon you will begin to detect movement. The swiftest animals in the pool will probably be the small (two to three inches long) fish known as tidepool sculpins (*Clinocottus* spp. and other genera, Color Plate 16a). These fish will remain almost motionless on the bottom, and their mottled color pattern makes them difficult to detect. Suddenly they will dart out and then they can be followed.

Other rapidly moving, but less-often seen animals, are small shrimp like the red-banded shrimp, *Heptacarpus pictus* (Figure 6-23) which is found from San Francisco to Baja. These small shrimp (one inch long) can be quite numerous in lower pools. However, they vary in color, and often have red bands that break up their outline making them very difficult to see when still. The shrimps are quiet during mid-day, but start moving actively at dawn and dusk. They are omnivorous, and will feed on whatever they can scavenge or catch. Several species of these colorful shrimp occur in the intertidal zone up and down the California coast. To see these tidepool shrimp, carefully search any algae or surfgrass present in the pool; separate the blades slowly and watch for movement. Once discovered, you will be amazed at their speed and their number.

Hermit Crabs. The industrious hermit crabs (*Pagurus* spp., Figure 6-24) can be seen moving around the pool during the day and night. Hermits carry their protection with them in the form of an empty snail

Figure 6-24. A hermit crab, Pagurus *sp., emerges from its shell. These crabs are usually quite abundant in tidepools.*

shell into which they will retreat quickly if bothered. Otherwise, they scavenge about the pools in search of animal and plant debris that serves as their food— they are very effective scavengers! Hermit crabs have been described as the garbage men of the intertidal zone. Hermits will carefully inspect prospective shells, and only after careful scrutiny will they try one

Figure 6-25. Two-inch-tall brown turban snail, Tegula brunnea. Note the light-colored (orange) margin at the base of the foot.

Figure 6-26. Shell of a dire whelk, Searlesia dira, one and a half inches long. Note the small patch of light-colored encrusting alga near the anterior.

on for size. Large hermit crabs (half an inch or more in carapace length) are not above evicting smaller hermits from a shell they fancy.

In the spring and early summer, many very small hermit crabs can be found in these pools living in the small shells of the periwinkles, *Littorina* spp. Most of the larger hermit crabs will be housed in the empty shells of the black turban snail, *Tegula funebralis* (Figure 6-15), which may be found alive in the tidepools as well. Other hermits live in the shell of the brown turban snail, *Tegula brunnea* (Figure 6-25),

which occupies tidepools as well as other mid- and lower-intertidal habitats in northern and central California. In southern California the banded turban snail, *Tegula eiseni* (Figure 6-19), and the speckled turban snail, *Tegula gallina* (Figure 6-18), can also be found in these tidepools, and their shells are likewise used by hermit crabs.

Another gastropod mollusk (snail) found in the tidepools in central and northern California is the dire whelk, *Searlesia dira* (Figure 6-26). *Searlesia* is a carnivore and feeds on other snails and invertebrates such as barnacles. Its tall, spire-shaped shell is up to an inch and half in height, and is often encrusted with a pink or whitish coralline alga.

Tidepool Cnidarians. Many other carnivores occur in tidepools. The stationary hunters, the anemones, include the aggregating anemone, *Anthopleura elegantissima* (Figure 6-21), previously mentioned, and its much larger relative, the giant green sea anemone, *Anthopleura xanthogrammica* (Color Plate 5a). The green anemone will also be found in lower tidepools, and reaches its largest size in the low intertidal surge channels.

The striking proliferating anemone, *Epiactis prolifera* (Color Plate 5b), a smaller anemone (one inch in diameter) delicately patterned with white lines, may also be found in tidepools. *Epiactis* ranges in color from cherry red to orange to bright green. Compared to the other anemones that attach only to the solid bedrock, *Epiactis* frequently may be found attached to small stones, shells or even a blade of alga. The red animal in Color Plate 5b has several smaller anemones attached to its column. These are sexually produced young that are brooded by the female. When they grow larger, they crawl off mom's column and take up an independent existence.

Tidepool Sea Stars. Motile predators include the sea stars. Some sea star species have a truly intertidal distribution, while others appear to forage into the intertidal zone from below during high tide and retreat to tidepools when the tide recedes. The small (up to four and a half inches across), six-rayed sea star, *Leptasterias* spp. (Figure 6-27), appears to remain in the intertidal zone and often shows up in tidepools where it searches for small molluscan prey. The six-rayed star is common in northern and central California, where at least two species' distributions overlap. These species are difficult to distinguish and I have elected to lump them as *Leptasterias* spp.

The large (up to eight inches across) leather star, *Dermasterias imbricata* (Figure 6-22), may also be found in tidepools. The leather star is smooth, slippery and blue-gray in color with red or orange mottling. Leather stars prey on sea anemones including *Epiactis prolifera* and the two *Anthopleura* species. The leather star is found as far south as San Diego,

but is uncommon in the intertidal zone in southern California.

The Pacific sea star, *Pisaster ochraceus* (Color Plate 14b), is a generalized predator that will feed on a wide variety of invertebrate prey. Although capable of withstanding complete exposure during low tide, *Pisaster ochraceus* will also show up in tidepools, and therefore deserves mention here. The Pacific sea star is the most common, large sea star in the middle and lower intertidal zones from Alaska to Santa Barbara. It is also abundant in mussel clumps. This sea star reaches a diameter of over 18 inches, although usually smaller, and varies in color from yellow or pale orange to dark brown or deep purple. It has many small, white spines on its aboral surface (side opposite of the mouth), which form a pentagonal pattern on the central disk. *Pisaster ochraceus*' role in the mussel clump is described later in this chapter. It also feeds on a variety of other prey including *Tegula* spp., chitons, barnacles, and limpets.

The sea bat, *Patiria miniata* (Figure 6-28), is also a common tidepool inhabitant. The sea bat receives its name from its five, stout, triangular arms that appear almost webbed, like the extended wing of a bat. This star reaches a diameter of eight inches, and varies in color, most commonly orange or light red, and is frequently mottled with other colors. Sea bats will eat almost anything, and are capable of everting (turning inside out) a large stomach over their plant and ani-

Figure 6-27. Two six-rayed sea stars, Leptasterias *sp., against a backdrop of encrusting coralline algae. Stars are one and half inches across.*

Figure 6-28. The omnivorous sea star, Patiria miniata, *commonly known as the sea bat.*

mal food, dead or alive. In addition to almost any animal food, sea bats feed on surfgrass (probably on the plants growing on the surfgrass, called epiphytes) and algae, as well as the organic film on rock surfaces.

The sunflower star, *Pycnopodia helianthoides* (Color Plate 14a), sometimes shows up in the upper tide pools, although it is uncommon in the intertidal zone south of Carmel Bay in central California. This beautiful, multi-rayed sea star will sometimes seek shelter in an upper pool when trapped by a receding tide. It is the largest and fastest moving of all our sea stars and individuals well over 40 inches have been described. Sunflower stars feed on a variety of prey but prefer purple sea urchins, *Strongylocentrotus purpuratus* (Color Plate 15a). These urchins occur occasionally in upper pools, but are much more common in tidepools in the low intertidal zone. In addition to urchins, this sea star will also take mussels, chitons, snails, and other sea stars such as the smaller *Leptasterias*. Sunflower stars have twenty or more arms, and they are usually purple in color, although an occasional yellow-orange individual is seen. Compared to the rigid-bodied Pacific sea star, *Pisaster ochraceus*, *Pycnopodia* is quite soft and flimsy. Improper or prolonged handling will cause this sea star to lose its arms, so it is best left at rest when found.

Under Tidepool Rocks

Under-Rock Crabs. If a mid-level pool is fairly deep and contains small boulders and a little sediment, another group of animals may occur. Remember to be careful when turning rocks over anywhere in the intertidal environment, and always carefully replace rocks the way you found them. Most prominent in this under-rock niche can be the red rock crab, *Cancer antennarius* (Figure 6-29). This crab may be quite large, over four inches across the carapace (back), although larger specimens are usually found in the low intertidal zone (see also Figure 7-8). The juvenile specimen illustrated here is more typical of the middle zone under rock habitat. It shows the prominent antennae that give this crab its scientific name "*antennarius.*" The rock crab has black-tipped claws that it uses for crushing the shells of its prey, mollusks and hermit crabs. Specimens of the red rock crab are typically found under rocks, and burrowed into the sediment with only a portion of their carapace showing. During night time high tides they emerge and prowl the bottom.

Another crab in this under-rock hiding place is the small, black-clawed pebble crab, *Lophopanopeus bellus*. (Figure 6-30). The male black-clawed crab reaches a carapace width of one and a third inches. Its body is usually tan to dun colored, although quite variable in color, and the carapace and walking legs are hairy. The claws are quite large in proportion to

Figure 6-29. A juvenile rock crab, Cancer antennarius. *Note the long antennae and the black-tipped claws.*

Figure 6-30. The black-clawed pebble crab, Lophopanopeus bellus. *Several species of these small crabs are found in California.*

the walking legs, and are black at the tips. This crab scavenges for its food, eating a variety of plant and animal matter. The black-clawed crab is found from Alaska to San Luis Obispo County in central California. There are several other species of *Lophopanopeus* found along the entire California coast.

If there is a considerable layer of sandy gravel in a tidepool, you might find a small burrower, the beach ghost shrimp, *Callianassa affinis.* This small (two and a half inches long) animal looks very similar to the ghost shrimp described in the mud flat section (Figure 5-26; Color Plate 13b), and feeds on detritus like its larger kin. The beach ghost shrimp occurs in mated pairs and is found south of Santa Barbara.

Brittle Stars. A small, delicate brittle star, *Amphipholis squamata* (Figure 6-31), and the similar-appearing, but much longer-armed, *Amphiodia occidentalis,* are common under rocks that lie on top of sand and fine gravel in tidepools. They may also occur among the holdfasts of large algae. Typically, if one brittle star is discovered, careful inspection will turn up a number of these animals. Both species mentioned here feed on suspended and deposited organic material, and will autotomize (drop off) their arms if handled. They are best observed without being picked up. These brittle stars range in size from one half to three inches in diameter.

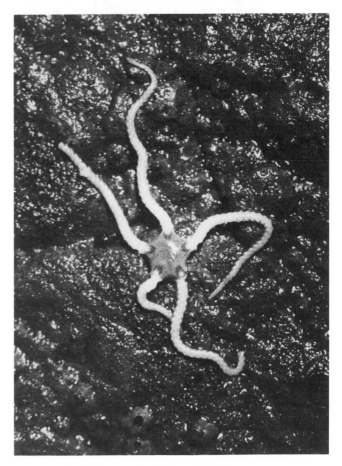

Figure 6-31. The small, delicate brittle star, Amphipholis squamata. *This echinoderm can be quite abundant under rocks.*

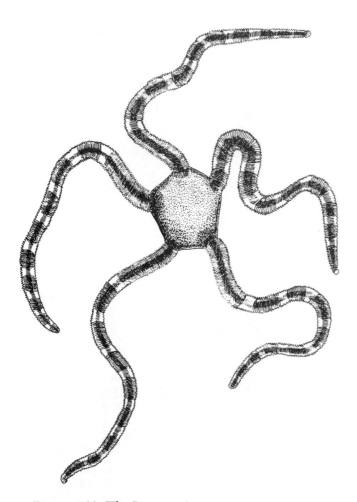

Figure 6-32. The Panamanian serpent star, Ophioderma panamense. *This large serpent star can reach a foot or more in diameter.*

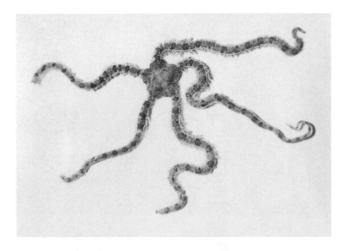

Figure 6-33. Under rock brittle star, Ophionereis annulata. *Note the conspicuous bands on the arms.*

Other, larger brittle star species occur under rocks in southern California, including the large Panamanian serpent star, *Ophioderma panamense* (Figure 6-32). This serpent star reaches a diameter of over a foot, and is a nocturnal carnivore and scavenger. The central disk is circular in outline and an inch in diameter. The disk is a velvety olive green, and the gray-brown arms are banded in white.

Another common southern California brittle star is *Ophionereis annulata* (Figure 6-33), which extends south from San Pedro in Los Angeles County to Panama. This star has a half inch wide brown disk, and arms that reach four to five inches in length. The slender arms are gray, marked with brown rings and flanked by stout, short spines. This brittle star is truly "brittle," and will readily shed all or pieces of its arms if handled. This animal moves quite fast when exposed, and is best left alone to continue its scavenging for bottom detritus.

Under-Rock Fish. Common under-rock denizens of tidepools are the small fishes with elongate, eel body forms. There are several types of eel-like fishes that occur including gunnels, snake eels, snailfishes, blennies, and pricklebacks [3]. These small, under-rock individuals are generally two to five inches in length. Figure 6-34 shows a prickleback, *Xiphister* sp. Larger specimens move downward in the intertidal and into the subtidal. Pricklebacks feed primarily on algae with some animal material in their diets. They will also be found under exposed rocks throughout the middle and low intertidal zones, as will most other under-rock, tidepool species mentioned here.

Another interesting fish found occasionally under tidepool rocks is the clingfish, *Gobiesox* sp. (Figure 6-35). The northern clingfish, *G. meandricus*, reaches six inches in length and is common in northern and central California. The smaller (three inches long) California clingfish, *G. rhessodon*, is found from Pismo Beach in San Luis Obispo County south to Baja. As their name implies, clingfishes cling to the rock with pelvic fins modified to function like a large suction cup. Clingfish will attach to the palm of your hand, and can support their own weight when inverted. Seen from above, the clingfish looks like a tadpole or a frying pan with its broad, flat head and small tapering body. Clingfishes feed mainly on small crustaceans. Like the prickleback, clingfishes may be found under exposed rocks in the mid and low intertidal zones.

Figure 6-34. A small (three inches long) prickleback eel, Xiphister *sp., viewed from above. Note the dark facial stripes and mottled body markings.*

Figure 6-35. A northern clingfish, Gobiesox meandricus, *three inches long. Note the large flat head and tapering body.*

Rock Undersides. When turning rocks in tidepools to discover what lurks on the substrate underneath, don't forget to look at the bottom of the rock itself. Here you will find a very interesting array of animals ranging from the juveniles of sea urchins and abalones to a plethora of small mollusks and worms. They are too numerous to try and cover all of them in this guide, but I will mention some of the more prominent. Again, remember to be careful while replacing the rock the same way you found it.

First, look closely at the bottom of the rock, and watch for movement. The most active animals will

probably be small (half inch or less) amphipod crustaceans (Figure 6-36), which can occur in bewildering numbers of many species. Also common can be isopods such as *Cirolana harfordi* (Figure 6-37). The amphipods are primarily algae feeders, while *Cirolana* is a scavenger on almost any animal tissue.

A tan to pale gray, one-to-two-inch-long flatworm, *Notoplana acticola* (Figure 6-38), is common here, and can be seen gliding along smoothly on a carpet of cilia. Look for a pair of eyespots near the front end of the worm. These very flat worms are perfectly suited to the tight quarters of the under-rock habitat. They feed on small mollusks and crustaceans, and range more widely than the under-rock niche at night. Once you have discovered one flatworm, chances are excellent that there are several more close by. They are not limited to tidepools, virtually any rock resting on a slight bit of moist sediment in the upper, middle, or low intertidal zones may harbor flatworms underneath.

The bottoms of rocks can also harbor various polychaete worms. Look closely for very small (a tenth of an inch or less), white, spiral shells (Figure 6-39; see also Figure 7-20). These are made by a worm in the genus *Spirorbis*. There are several species of these worms and their identification is confusing. Not only do they attach to the undersides of rocks, but also to algae, shells, and even the skeletons of crabs. The worm itself protrudes a bright red ring of tentacles for filter-feeding when submerged. One of the tentacles is

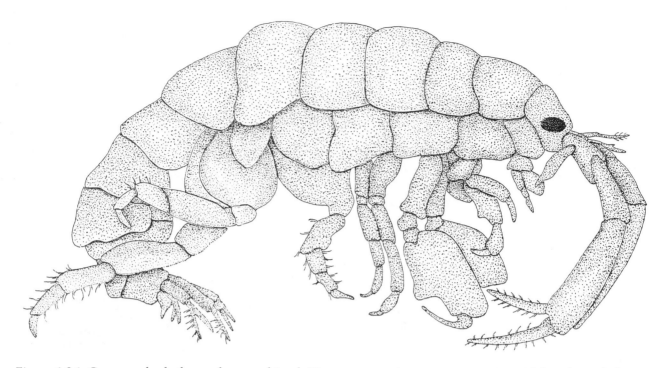

Figure 6-36. Common body form of an amphipod. Numerous species occur among intertidal rocks and algae.

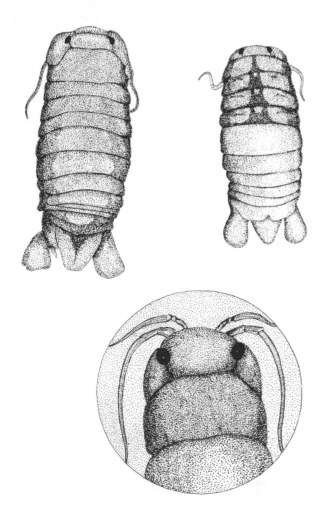

Figure 6-37 (Left). The scavenging isopod, Cirolana harfordi, *common under intertidal rocks.*

modified into a stopper-like operculum to close off the shell. Once you have identified this worm you will start seeing it everywhere!

Less abundant, but much larger will be the thin, membranous, sand-encrusted tubes of the tentacle-feeding worms of the polychaete family Terebellidae. These worms are closely related to the large worm *Pista pacifica* (Figure 5-5) of the sand flat, introduced in Chapter 5. Figure 6-40 shows *Thelepus crispus*, which can be almost a foot long, although smaller individuals are usually seen. It remains in the safety of its semi-permanent tube and sends forth long, off-white, prehensile tentacles that creep along on cilia to scour the bottom of the tidepool for deposited organic material. These tentacles emerge from the worm's head region along with bright red gills used for breathing.

If there is a substantial amount of fleshy algae in a pool, searching through it will usually produce a swarm of small amphipod crustaceans and several small snails. A larger find (one to two inches across the back) may be the spidery-looking kelp crab, *Pugettia producta* (Figure 6-41). Also called the

Figure 6-38. One-inch-long intertidal zone flatworm, Notoplana acticola.

Figure 6-39 (Below, Left). Feeding tentacles and operculum emerge from the small (one tenth of an inch diameter) spiral shell of the polychaete Spirorbis sp.

Figure 6-40 (Right). The polychaete worm, Thelepus crispus, spreads its feeding tentacles. This worm has been removed from its heavy mucous tube.

Figure 6-41 (Below, Right). The shield-back or kelp crab, Pugettia producta, one and a half inches across the back. Note the spidery, pointed legs.

shield-backed crab, *Pugettia producta* ranges in color from green to dark brown and blends with its algal cover. The young kelp crab feeds on algae, and as it grows it moves downward in the intertidal zone and takes on a more carnivorous diet. Kelp crabs can also be found in several lower intertidal and subtidal habitats, and can reach a size of over three and half inches across the carapace (back).

MIDDLE INTERTIDAL ZONE

The middle intertidal zone extends from two and half feet above MLLW to zero MLLW. It is the part of the intertidal that is exposed by most low tides and covered by most high tides, so it spends about half the time submerged and half exposed. Unlike the upper and high intertidal zones that have relatively low diversity, the middle intertidal is awash with life. We have already discussed tidepools that occur in this zone, and there are several other fairly discrete habitats to consider. The rest of this chapter discusses the exposed-rock and mussel clump habitats of the middle intertidal zone.

Exposed-Rock Habitat

Often, portions of the rocky intertidal reef rise up above the flat bedrock of the reef face. These upraised areas are called outcroppings, and they provide rocky surfaces that are exposed to desiccation (drying out) more than adjacent flat reef areas (Figure 6-42). Similarly, large boulders that remain relatively stationary also provide this elevated habitat. In relatively unprotected intertidal areas that are exposed to considerable wave action, this upper exposed-rock surface habitat grades into the mussel clump habitat as subtle changes in tidal height and the degree of wave exposure occur. Thus many species occur in both habitats. In general the exposed-rock habitat is higher in tidal elevation and thus more subject to drying than is the mussel clump. The main line of demarcation between these two habitats represents the upper physiological limit of the California sea mussel, *Mytilus californianus,* which provides the superstructure of the mussel clump habitat discussed later.

Exposed-Rock Seaweeds. The algae of the exposed rock habitat are similar to those found in the upper tidal zone. Tufts of coarse, scrubby alga, *Endocladia*

Figure 6-42. An exposed vertical rock face. Note the limpets in the upper zone, the mussels and barnacles at mid zone, and algae beneath.

Figure 6-43. Small tar spot algae, Ralfsia sp., growing on an exposed rock surface.

muricata, are found on the tops of outcrops, and rock-weeds, *Fucus* and *Pelvetia*, hang down from the sides. Other algae include the thin sea lettuce, *Ulva* sp., and the red alga known as brown laver or nori, *Porphyra* spp. (Color Plate 2c), both of which can be quite abundant in the late spring and summer. On the lower, vertical surfaces of outcrops several species of fleshy red algae, and a few large brown algae known as kelps, grow in profusion, especially in summer.

A final brown alga to look for is the tar spot alga *Ralfsia* spp. (Figure 6-43). As the common name implies, *Ralfsia* looks like a dried tar spot on the rock. Beachcombers may think they've found the aftermath of an oil spill! The alga starts out as a circular, dark brown to black crust and then grows outward into patches up to eight inches in diameter. It may have slightly raised concentric ridges or be flat. *Ralfsia* is

highly resistant to drying, and consists of short, erect, tightly-packed plant filaments that grow so close together that they form a solid tissue.

Barnacles. The most abundant animal found on these exposed rocks often is the volcano-shaped, acorn barnacle. Barnacles are filter-feeders and are active when they are covered with water. When seen at low tide they are closed up and inconspicuous. The small barnacles found highest on the outcrop are *Balanus glandula* and *Chthamalus* sp. (Figure 6-9), the most desiccation-resistant of the common barnacles and already introduced in the upper tidal zone. In northern and central California a larger species, *Balanus cariosus* (Figure 6-44), is found lower on the sides of the outcrop mixed with *Balanus glandula* and *Chthamalus*. This barnacle is common on exposed

Figure 6-44. Large acorn barnacle, Balanus cariosus, *of the middle intertidal zone. Smaller barnacles are* Chthamalus *sp.*

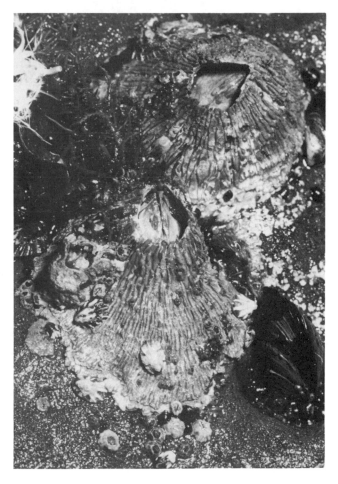

Figure 6-45. Two volcano-shaped Tetraclita rubescens, *a pink-tinged barnacle that reaches two inches in diameter.*

rocks and on mussels. It grows to a larger size (up to two inches or more in diameter) than the smaller, more abundant barnacles, has a thatched or ridged appearance, and is gray in color. *Balanus cariosus* usually occur as individuals rather than in a large aggregations like *B. glandula* and *Chthamalus*. If it escapes predation by carnivorous gastropods and sea stars, *Balanus cariosus* can live up to 10 to 15 years.

A fourth species of barnacle, *Tetraclita rubescens* (Figure 6-45), is also represented here, and occurs from Tomales Bay in Sonoma County south to Baja California. *Tetraclita* also has a thatched or ridged shell similar to *Balanus carious*, but has a distinctive pinkish-red color and has a smaller opening. This barnacle is usually an inch or so in diameter, although two-inch specimens occur. It is most commonly seen in shady places on the outcrop, especially on the "ceiling" and sides of overhangs, caves, and ledges. These latter, shady surfaces are typically shared with pale-colored aggregating anemones, *Anthopleura elegantissima*. A fifth barnacle species, the stalked barnacle *Pollicipes polymerus,* will also be found in sheltered cracks on the outcrop, but reaches its peak abundance in mussel clumps.

Exposed-Rock Snails. In the lower, more shaded portion of the exposed-rock habitat the abundant, stationary acorn barnacles are ready prey for several carnivorous snails. In northern and central California the most obvious predator is the snail known as the emarginate dogwinkle or whelk, *Nucella emarginata* (Figure 6-46). It occurs high up on these rocks and nestles among its prey during low tide. Like the other barnacle predators, the dogwinkle is a "drill," and uses its radula to scrape or "drill" holes into the shell of its prey and get at the soft, edible parts. Careful inspection of the barnacles nearby these snails will reveal that many are the empty shells of past victims. *Nucella* has a compact, heavy shell with a low spire. The shell can reach over an inch and a half in height, but is usually smaller. Colors vary from dark brown,

Figure 6-46. Two half-inch specimens of the emarginate dog winkle, Nucella emarginata. *Shell pattern is highly variable in this snail.*

Figure 6-47. Shells of the frilled dogwinkle, Nucella lamellosa, *showing the variation in shape and sculpturing.*

Figure 6-48. Two rock snails, Ocenebra circumtextra, *three quarters of an inch long. Specimen on left shows the banding pattern characteristic of the species.*

gray, or black to an occasional orange individual; the interior of the shell is purple. *Nucella* feeds mainly on barnacles and mussels, although it will also prey on *Tegula funebralis*, *Littorina scutulata*, and *Collisella scabra*.

In northern California another, larger dogwinkle species, *Nucella lamellosa* (Figure 6-47), can be found. This species is known as the frilled dogwinkle or wrinkled purple, and can reach up to two inches in height. As the common names imply, the shell of this species can be highly ornamented with frilled ridges and spiral bands. However, it can also be nearly smooth. The thinner, ornamented shells are found on snails in habitats where there are few predators, while the heavy, smooth shells occur on animals living in situations where predation, chiefly by large crabs, is common. The shell color varies from white to orange to brown, and can be a solid color or banded. In northern California you are likely to find this snail somewhat below the level of the emarginate dogwinkle. In central California this species seldom occurs in the intertidal zone, although its shell is occasionally found washed up on the beach or on a hermit crab. Like the emarginate dogwinkle, this species feeds chiefly on barnacles and mussels.

A smaller, banded snail, *Ocenebra circumtextra* (Figure 6-48), also feeds on barnacles on the exposed rocks, and is commonly found in areas of heavy surf. The shell is low-spired, reaches a length of one inch,

Figure 6-49. An inch-long barnacle drill, Acanthina spirata, *showing the diagnostic checkered patten and stout tooth on the shell's aperture.*

although usually smaller. It is gray with a distinct pair of brownish bands on each shell whorl (each complete "wrap" of the spiraling gastropod shell is called a whorl). Their chief prey appear to be barnacles, although other small prey, such as mussels and snails, are probably also eaten.

The angular unicorn snail, *Acanthina spirata* (Figure 6-49), is another barnacle predator, and is recognized by a stout tooth near the opening (aperture) of its shell. *Acanthina spirata* occurs on moderately exposed shores from Tomales Bay in Sonoma County to Baja California. It is also common on the rocks inside coastal embayments and on breakwaters. This snail can reach over an inch and a half in height. The unicorn snail forms mating aggregations in the spring and early summer, and the females lay their eggs into flask-shaped capsules. Clumps of these aggregating snails and their capsules can be found under large boulders in the middle intertidal exposed rock habitat. The young snails emerge from these capsules ready to take up the adult lifestyle. Their main predator is the six-rayed sea star, *Leptasterias* spp. (Figure 6-27).

Non-predatory snails found here include the ever-present black turban, *Tegula funebralis* (Figure 6-15). This snail sometimes aggregates in groups of hundreds on the protected, shoreward sides of outcrops. On outcroppings high in the intertidal, the periwinkle, *Littorina scutulata* is common.

Limpets and Black Abalone. The ribbed and digit limpets, *Collisella scabra* and *C. digitalis,* are also found on the outcropping, taking advantage of the microalgae growing there. The larger shield limpet, *Collisella pelta* (Figure 6-50) can also occur here, usually under the protective covering of some overhanging alga. It is a robust limpet, reaching over an inch and a half long, and highly variable in color, ranging from green to black and often with white checks or rays. The shell's apex (top) is near the center of the shell, and the sides are all convex and frequently ribbed. As is the case with many larger limpet species, the top of the shell is often eroded. Shield limpets feed on a variety of fleshy brown and red algae including *Endocladia* (Color Plate 2d), *Iridaea* (Color Plate 3d), *Egregia* (Figure 6-67), and *Postelsia* (Figure 6-66). This limpet shows a marked escape response to a number of intertidal sea stars that are its chief predators. It lifts its shell well off the substrate and very quickly (for a limpet) crawls away.

A limpet frequently found near the shield limpet on semi-protected rocks is the file limpet, *Collisella limatula* (Figure 6-51). The file limpet is so named for the strong, prickly radial ribs that grace its low profile shell. This limpet reaches over an inch and three quarters long, and the shell is buff or yellowish or greenish brown in color, sometimes with dark mottling or white spots. It feeds primarily on microscopic algae, and algae that form sheets like the pinkish encrusting coralline algae (Color Plate 2a).

Figure 6-50. The shield limpet, Collisella pelta, *two inches long. Note the worn robust shell.*

Figure 6-51. The file limpet, Collisella limatula, *showing the characteristic shell sculpturing of this species.*

In northern and central California, a fifth limpet species, the plate limpet *Notoacmea scutum* (Figure 6-52), may be found closer to the base of the outcropping. This limpet is quite flat and broad and often sports tassels of attached algae. The plate limpet's shell reaches almost two and a half inches in length, is low in profile, and the apex is round and near the center. These characteristics give the limpet a flat, plate-like appearance that distinguishes it from the other limpet species. Like the file limpet, plate limpets are grazers on microscopic algae and on encrusting coralline algae. They are preyed upon by the green-lined shore crab, *Pachygrapsus crassipes* (Figure 6-8), and by the sea stars *Leptasterias* (Figure 6-27), *Pisas-*

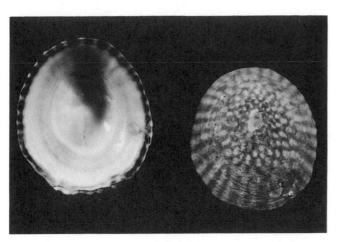

Figure 6-52. Top and bottom views of the large, flat shell of the plate limpet, Notoacmea scutum.

Figure 6-53. A two-and-a-half-inch-long owl limpet, Lottia gigantea, *rests on its home scar in a territory surrounded by mussels and barnacles.*

ter (Color Plate 14b), and *Pycnopodia*. The plate limpet has a running escape response that is elicited by these predatory sea stars. This response is similar to that described for *Collisella pelta* (Figure 6-50).

The "granddaddy" limpet of them all is the owl limpet, *Lottia gigantea* (Figure 6-53). *Lottia* occurs on the most exposed middle intertidal rocks. This is the largest of our limpets, with a shell over three and a half inches in length. The shell is low in profile and the apex is near the anterior margin (front edge). The outer surface is rough and eroded, and brown with whitish spots. At low tide large individuals of this species occupy "home scars" on the rocks that fit the margin of their shell. Studies show that each limpet lives within a "territory" approximately the size of a dinner plate, grazing on the algal film that grows there. Striation marks in the center of Figure 6-53 are grazing scrapes left by *Lottia*'s radula. Each *Lottia* remains in its territory and keeps it free from other animals that may move in from adjacent areas. Interestingly enough, the shell of *Lottia* often constitutes the home scar of the smaller rough limpet, *Collisella scabra* (Figure 6-5). *Lottia* is preyed upon by shorebirds such as oyster catchers and occasionally by humans.

A final herbivorous snail should be mentioned here. This is the black abalone, *Haliotis cracherodii* (Figure 6-54). Black abalone are seen occasionally in crevices on outcroppings. They also turn up under rocks in the middle and low intertidal zones. They are more common south of Point Conception. The black abalone is so named for its dark-colored shell, which can be dark blue, dark green, or black. The shell is usually smooth and lacks the growth of encrusting marine plants and animals typically found on the shells of other abalone species. Historically, black abalone were quite numerous throughout the intertidal zone, occurring in the open on even the most exposed rocks like the limpets previously mentioned. They can still be seen in these habitats on the California Channel Islands which have restricted access. On mainland California human predators have reduced their numbers, so that now they are found only in well-hidden niches. Black abalone can reach a shell length of eight

Figure 6-54. A black abalone, Haliotis cracherodii, *four inches long. Note the relatively clean shell on which a small limpet has excavated a home scar.*

inches, but usually only smaller individuals are seen. These animals feed chiefly on drifting algae, especially the large brown algae or kelps. Black abalone are preyed upon by sea otters, octopuses, the sea star *Pisaster ochraceus* (Color Plate 14b), and by fishes.

Exposed-Rock Crabs. The purple shore crab, *Hemigrapsus nudus* (Figure 6-55), and more commonly, the green-lined shore crab, *Pachygrapsus crassipes* (Figure 6-8), can be found in this exposed-rock subhabitat. These crabs can be seen holed up in cracks and crevices, and the green-lined shore crab is often seen scuttling all about the rock surface. The purple shore crab is more abundant in undercut or overhanging areas of outcroppings, which it shares with *Tetraclita* (Figure 6-45) and aggregating anemones. Large male purple shore crabs can reach two inches across the carapace (back) although most are smaller. The claws are marked with reddish-purple spots and there may be some purple on the dark brown to red body. Shore crabs feed on low-growing algae that they crop with their claws, and scavenge for animal remains. Although the purple shore crab is known to occur from Alaska to Baja California, they are not commonly seen in southern California. Their chief predators are shorebirds and fish.

Sand Castle Worms. A remarkable worm colony can sometimes be found at the base of upraised, exposed rocks in the middle and low intertidal zones. These are the homes of the sand castle worm, *Phrag-*

Figure 6-55. A male purple shore crab, Hemigrapsus mudus, *two inches across the carapace. Note the prominent dark (purple) dots on the chelipeds.*

matopoma californica (Color Plate 7c). Sand castle worms construct their tubes by cementing together individual sand grains. The resulting colony has a distinct honeycombed appearance and can be yards across. The worm itself is only an inch long, and protrudes a circle of stiff tentacles out of the tube's opening during high tide to filter feed. These worms can be locally common from central California south. They need a local source of sand for their tubes, so they are found more predictably in rocky intertidal habitat that has a sandy beach nearby.

Mussel Clump Habitat

Sea Mussel. Large beds or clumps of the California sea mussel, *Mytilus californianus* (Figure 6-56), are common all along the open California coast. The mussel thrives in areas of high wave energy and, indeed, its distribution corresponds to the most exposed, wave-tossed rocky intertidal habitat. Mussels reach five inches or longer in length in the intertidal zone, and up to ten inches in subtidal clumps. The shell is thick, pointed at the anterior end, and sculptured with radiating ribs and growth lines. The shell is often eroded or worn from abrasion in the clump, and frequently colonized by barnacles and limpets (Figure 6-57). Mussels are filter-feeders, and the average mussel filters two to three quarts of sea water an hour when submerged. They filter mainly small particulate detritus and dinoflagellates.

One dinoflagellate commonly filtered is *Gonyaulax catanella*, which produces a toxin that can cause paralytic shellfish poisoning in humans (see Chapter 2). Mussels may accumulate large amounts of the toxin during summer months when the dinoflagellates can be most abundant. Therefore, the mussels are quarantined in California between May 1 and October 31. It is always a good idea to check with local authorities when taking any shellfish in California, as sporadic plankton blooms of dinoflagellates can occur even during the "safe" season, and shellfish poisoning is a real possibility.

Mytilus californianus is able to settle and grow in wave exposed situations because it attaches itself to the bedrock with stout fibers made of protein called byssal threads. In time the mussel can monopolize the horizontal surfaces to the exclusion of other species that require attachment to the rocky substrate. How-

Figure 6-56. *A clump of the California sea mussel,* Mytilus californianus, *an animal at home in the exposed middle intertidal zone.*

Figure 6-57. *A large California sea mussel,* Mytilus californianus, *serves as a host to barnacles and a limpet.*

ever, mussels are not the only organisms that can inhabit this habitat successfully. First of all, the sea mussel is ultimately its own worst enemy. As the mussels grow and new larval mussels recruit into the mussel clump, the clump becomes more and more unstable. This occurs because the clump becomes several mussels deep, and fewer mussels are attached directly to the bedrock. Instead, they are attached to one another by their byssal threads. The growing clump offers more and more resistance to the pounding waves with less and less direct attachment, until large areas of the clump are torn away, leaving bare rock behind. This rock becomes available for new settlement, and of new organisms move in. Thus begins a cycle that may take seven or more years and will end up with *Mytilus* back in control only to grow and be torn away again.

Role of the Pacific Sea Star. The mussel's upper distribution is limited by its physiological tolerance to exposure. In areas of consistent wave action, the mussels can inhabit high intertidal zones successfully because of the wave splash, but in areas of periodic calm, their upper limit is the middle intertidal zone. Another control of the mussel is the Pacific sea star, *Pisaster ochraceus* (Color Plate 14b). *Pisaster*'s predation on the mussel is sufficient to preclude it from moving into and monopolizing the lower intertidal zone in the way it can dominate the middle intertidal zone. Thus, *Pisaster* keeps the lower substrate open for other species to colonize and inhabit, allowing for a much more diverse assemblage of organisms in the mussel clump habitat.

In addition to *Pisaster* and *Mytilus*, several other prominent organisms inhabit the mussel clump. The

Figure 6-58. A stalked barnacle, Pollicipes polymerus, *with its filtering appendages in the normal feeding position.*

stalked barnacle, *Pollicipes polymerus* (Figure 6-58), occurs in round aggregations, sometimes surrounded by mussels. This filter-feeder is capable of rotating its upper body on its muscular fleshy stalk. It can thus position its filtering appendages, a series of modified legs with closely placed filtering hairs, into the current for the most favorable feeding on large particles of detritus and large zooplankton. The individual stalked barnacle may reach a length of three inches. Close inspection may reveal smaller barnacles attached to its stalk, because like the mussel, *Pollicipes'* larvae seek out and settle on the adult animals. Large barnacles are relatively immune to attack by most intertidal predators. Look closely at the photo of the *Pollicipes'* clump (Figure 6-59) and you will spot two cleverly disguised color variants of the digit limpet, *Collisella digitalis.* One is just to the left of center and the other is below center. Find them? (Look at Figure 6-4 to refresh your memory of *Collisella.*)

The solitary giant owl limpet, *Lottia gigantea* (Figure 6-53), is a conspicuous loner compared to the "togetherness" of *Mytilus* and *Pollicipes.* Individual *Lottia* will actively maintain territories in the middle of a dense mussel clump. As previously discussed, the territories consist of patches of closely cropped algae on which they graze. *Lottia* discourages intrusions onto its territory by other animals, including mussels and barnacles, through a series of very specific protective behaviors. Each *Lottia* retreats to a distinct home scar within its territory at low tide.

Figure 6-59. A colony of the stalked barnacle, Pollicipes polymerus, *with two camouflaged digit limpets,* Collisella digitalis.

Figure 6-60. Nuttall's chiton, Nuttallina californica, *two and a half inches long. Chiton is nestled in a rocky depression among acorn barnacles.*

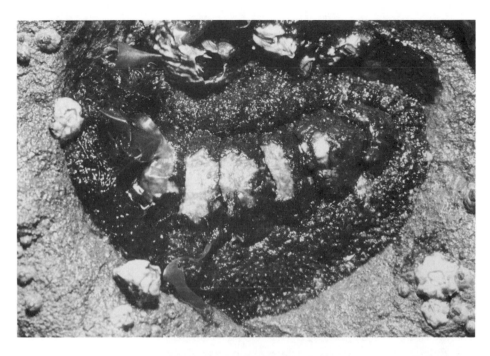

Another mollusk that lives on a home scar among the mussels or near the edge of the clump is Nuttall's chiton, *Nuttallina californica* (Figure 6-60). *Nuttallina* leaves its home scar at high tide to graze nearby algae and returns with the ebbing tide. It prefers erect coralline algae, but also eats *Endocladia* and *Cladophora.* The chiton's body is up to two inches long, and the shell plates are dark brown, although usually eroded or covered with algal growth. The girdle is dull-colored and frequently overgrown by algae. The girdle is wide, coming up high on the shell plates, and is covered with short, rigid spines. This chiton's chief predator appears to be sea gulls.

The katy chiton, *Katharina tunicata* (Figure 6-61) is a second, larger (to five inches long) chiton also seen near mussel clumps. This chiton is robust and oval in shape. All the valves are deeply imbedded in a shiny black girdle. Only the very centers of the valves show, and these are often fouled with algae or animals. The bottom of the foot is a burnt orange color. Katy chitons feed on brown and red algae. They are not common south of Monterey Bay.

The surfaces of the mussels' shells also serve as available habitat (Figure 6-57). All four acorn barnacles mentioned in the exposed-rock subhabitat can be found on the mussels. *Tetraclita* (Figure 6-45) is particularly abundant. Numerous small limpets of several species including *Collisella scabra* (Figure 6-5), *C. digitalis* (Figure 6-4), *C. pelta* (Figure 6-50), and young *Lottia gigantea* (Figure 6-53) occur. Individu-

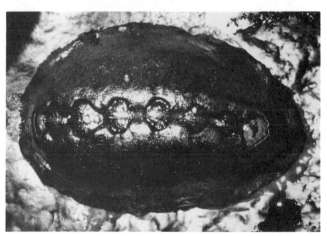

Figure 6-61. A four-inch-long katy chiton, Katharina tunicata. *Note how the shiny black girdle almost covers the shell plates.*

als of the aggregating anemone, *Anthopleura elegantissima* (Color Plates 4c and 4d), and the proliferating anemone, *Epiactis prolifera* (Color Plate 5b), can also be found on mussel shells.

With this abundance of small invertebrate prey, it is not surprising to find that the small, predatory six-rayed sea star, *Leptasterias* spp., and the barnacle-eating whelk, *Nucella emarginata* (Figure 6-46), are also common inhabitants of the mussel clump. These are only the larger, obvious animals of this habitat.

Organisms Within the Mussel Clump. Close scrutiny of the clump will reveal myriad smaller, motile animals and encrusting forms that combine to form one of the most diverse intertidal habitats. In addition to the shells of the mussels, the webs created by their collective attachment (byssal) fibers provide a maze of nooks and crannies for marine invertebrates to inhabit. Young sea urchins less than an inch across find shelter at the base of the clump. Worms are especially able to maneuver here, and a wide variety of species occurs. Prominent are the polychaetes known as clam worms, pile worms, or mussel worms, *Nereis vexillosa* (Figure 6-62), and the very similar *N. grubei*. *Nereis vexillosa* is common among mussel clumps and a variety of other intertidal habitats. *N. grubei*, which is also common among mussel clumps, inhabits seaweed holdfasts as well. These nereid worms are omnivorous, feeding on a broad diet of animal and plant tissues. They possess an elaborate, jawed, eversible proboscis that is used to capture live prey as well as tear off bits of algae. *N. grubei* reaches a length of four inches and is tan to gray colored; *N. vexillosa* is larger (to six inches long), and darker, often with iridescent highlights. These worms fall prey to *Cancer* crabs (Figure 4-21) and fishes.

Another polychaete frequently seen in the mussel clump is the scale worm, *Halasynda brevisetosa* (Figure 6-63). This worm reaches two inches in length and has 18 pairs of flat scales along its back. This scale worm is a scavenger and detritus feeder. It also occurs in the tubes of other large polychaetes like

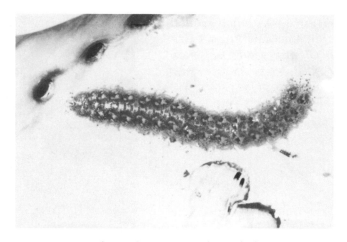

Figure 6-63. The scale worm, Halasynda brevisetosa, *a widely distributed errant polychaete.*

Figure 6-64. The nemertean worm, Emplectonema gracile. *This semi-contracted specimen is six inches long. Note the dark upper and creamy lower surfaces.*

Figure 6-62. The clam worm, Nereis vexillosa, *common in mussel clumps.*

Figure 6-65. Smooth porcelain crab, Petrolisthes *sp., three quarters of an inch wide. Note the large chelipeds (claws) and flattened body.*

Thelepus crispus (Figure 6-40), and *Pista pacifica* (Figure 5-5) of the sand flats.

The ribbon or rubber band worms (Phylum Nemertea, see Chapter 3) also thrive in the mussel clump. These elongated, predatory night-stalkers retreat into the protective maze of the mussel clump by day and actively forage through it and beyond at night. They are predators on a wide variety of invertebrates including polychaetes and small crustaceans such as amphipods and isopods. Several species occur here and they vary more by color than shape.

The most obvious ribbon worm is the bright orange *Tubulanus polymorphus* (Color Plate 6c). As the specific name *polymorphus* implies, this worm comes in a variety of colors, including yellow, vermilion and bright red, but orange is my favorite. It has a flattened, spade-like head and can reach a length of three yards when extended! It occurs from Alaska to San Luis Obispo County.

A second common species is *Emplectonema gracile* (Figure 6-64), which occurs in the mussel clumps, under rocks, in seaweed holdfasts, and in empty barnacle shells. This species is 4 to 6 inches when contracted, and can extend to 20 inches. It is yellowish green to dark green on the back, and pale greenish yellow to white on the bottom. *Emplectonema gracile* is often found huddled in tangled clusters. It also occurs in estuaries and on pier pilings.

One of the most common California ribbon worms is *Paranemertes peregrina,* shown with its everted proboscis clutching an isopod prey in Color Plate 6d. It is frequently seen among mussels and coralline algae and under rocks. It is up to six inches long contracted, and purplish brown, dark brown or orange brown on the back. Underneath and on the sides of its head it is deep yellow to white. This species stays hidden at high tide and ventures out in the early morning or at night at low tide. It is a predator of polychaetes and can swallow a worm bigger around than itself.

Crabs abound in the interstices of the mussel clump. Young of the familiar green-lined shore crab, *Pachygrapsus crassipes* (Figure 6-8), and the purple shore crab, *Hemigrapsus nudus* (Figure 6-55), are common. Other common crabs are the porcelain crabs, whose flat bodies allow them to move easily in tight quarters from which they seldom venture. Shown in Figure 6-65 is *Petrolisthes* sp. Four species of this genus occur along the California coastline and they all have this approximate shape. Porcelain crabs are so named because of their defensive behavior; when a leg or claw is grasped by a predator (or a human) they will cast off (autotomize) the appendage and run to safety. This drastic behavior led people to believe the crabs were delicate and easily damaged, thus the name "porcelain." Males reach about an inch or less across the carapace and the females are smaller. These crabs are filter-feeders, and use mouth parts (called maxillipeds) to set up filtering currents and strain plankton from the water. Porcelain crabs are preyed upon by other crabs, octopus, and fish.

Sea Palm and Other Mussel Clump Algae. An alga that is a prominent seasonal member of the mussel clump association from Morro Bay north is the sea palm, *Postelsia palmaeformis.* (Figure 6-66). This lovely kelp plant is a brown alga (Division Phaeophyta) although its color is green. Sea palms have hollow, flexible stems (stipes) that allow them to bend with the force of the waves and backwash, and snap back into their upright posture. Almost all sea palms are torn away during fall and winter storms but return, growing from microscopic germlings on the rocks, in the late spring.

Other, shorter-lived, seasonal algae of the mussel clump include the bright green sea lettuce, *Ulva* sp. (Color Plate 1a), and the thin brown laver, *Porphyra* sp. (Color Plate 2c). Also common here is the by now familiar coarse, scrubby alga, *Endocladia muricata* (Color Plate 2d). Studies have shown that newly settled mussels will initially attach to *Endocladia* and then reattach to the rock as they grow. This is one of the first steps in the mussels' recolonization cycle.

Another large brown alga that shows up here, and throughout the rest to the middle and low intertidal zones, is the feather-boa kelp, *Egregia menziesii* (Figure 6-67). This plant is also known as Venus' girdle.

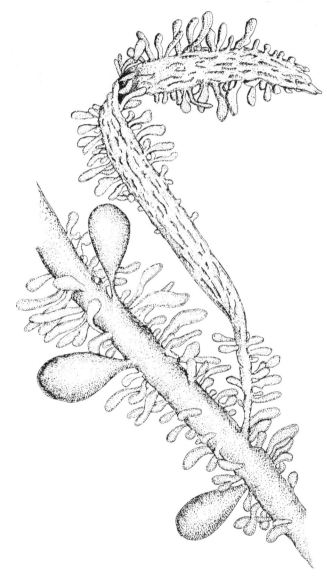

Figure 6-66. The sea palm, Postelsia palmaeformis, *a small kelp plant of the exposed middle intertidal zone.*

Figure 6-67. Central midrib and lateral blade of the feather-boa kelp, Egregia menziesii, *of the middle and low intertidal zones.*

It is a perennial plant that has a large, root-like hold-fast. From the holdfast strap-like stipes grow that can be several yards in length. The stipes consist of a stout, belt-like, midrib region from which grow short blades. Each blade has a small air bladder at its base that allows the kelp to float up off the bottom when submerged. During the winter the stipes break off, but the holdfast remains to produce more stipes in the spring. This plant harbors two very interesting hitch-hiking invertebrates.

Egregia's first hitchhiker is the large isopod, *Idotea stenops* (Figure 6-68), which occurs with the alga from Alaska to Point Conception. This animal is about an inch and a half long and matches the kelp's olive green to brown color perfectly. *Idotea stenops* feeds on the alga and is always found tightly clutched to the middle of the stipe. The second animal is the kelp limpet, *Notoacmea insessa* (Figure 6-69). The kelp limpet occurs all along the California coast and is less than an inch long. Its dark brown shell is thin, and the high sides are parallel and smooth. It also hugs the middle of the stipe from which it excavates depressions or scars with its radula. As can be seen in

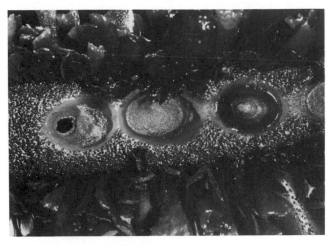

Figure 6-69. The kelp limpet, Notoacmea insessa, *residing in one of three depressions it has excavated along the midrib of* Egregia menziesii.

Figure 6-69, the limpet can excavate a series of these scars. They eat the alga and nestle into the depressions to prevent drying out during low tide. The limpet can move from stipe to stipe.

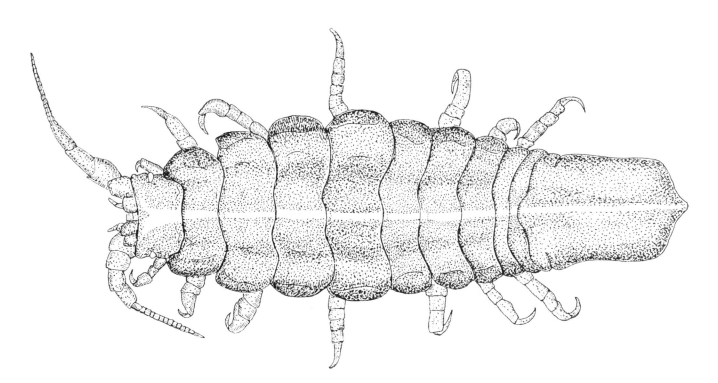

Figure 6-68. The isopod, Idotea stenops, *which lives attached to the midrib of the feather boa-kelp,* Egregia menziesii.

References

1. Ricketts, E. et al. *Between Pacific Tides*,. 5th ed. Stanford: Stanford University Press, 1985, 652 pp.
2. Brandon, J.L and F.J. Rokop. *Life Between the Tides: The Natural History of the Common Seashore Life of Southern California.* San Diego: American Southwest Publishing Company of San Diego. 1985, 228 pp.
3. Fitch, J.E. and R.J. Lavenberg. *Tidepool and Nearshore Fishes of California.* Berkeley: University of California Press, 1975, 156 pp.

7

The Rocky Intertidal Environment: Middle and Low Zones

❦

In many rocky intertidal habitats with a gradual slope, the middle and low intertidal zones can be quite extensive. In addition to the upraised rocks and outcrops discussed in the previous chapter, the middle and low intertidal zones can contain extensive stretches of flat, open space, and areas where small boulders have accumulated, called boulder fields. Both of these habitats grade relatively seamlessly between the middle and low intertidal zones (below 0.0 MLLW), and I will include the entire sweep of middle and low intertidal in this treatment of reef flat habitats. Finally, two remaining habitats, the low tide-pool and surge channel, are covered briefly.

REEF FLAT HABITAT

Algal Turfs

The flat, open areas of the middle intertidal zone are usually characterized by lawns of short-growing (two to four inches), red algae and are sometimes referred to as red algal "turfs" (Figure 7-1). The flat reef has little relief to break up the sweeping surf, and consequently the diversity of large invertebrates is relatively low here. The lush growth of red algae does provide ample fodder for herbivorous snails. The

Figure 7-1. Flat, wave-washed middle rocky intertidal habitat. Dark plants are red algal "turfs," lighter area is a patch of surfgrass, Phyllospadix *sp.*

black turban snail, *Tegula funebralis* (Figure 6-15), is very common and is often found with the small, hitch-hiking snails, the black limpet, *Collisella asmi* (Figure 6-16), and the slipper limpet, *Crepidula adunca* (Figure 6-17), attached to its shell. Lower on the reef the other turban snails can be found, including *Tegula brunnea,* in central and northern California, and *Tegula eiseni* (Figure 6-19) and *Tegula gallina* (Figure 6-18) in southern California. Hermit crabs, *Pagurus* spp. (Figure 6-24), usually in *Tegula* shells, are also very abundant on these reef flats. The dire whelk, *Searlesia dira* (Figure 6-26), is occasionally found among the algae, although it is more common in tidepools.

The dense growth of algae and their holdfasts trap moisture and wave-borne materials. The trapped moisture makes a hospitable setting for clones of the aggregating sea anemone, *Anthopleura elegantissima* (Color Plates 4c and 4D), which can be quite extensive. A layer of sand and small pebbles accumulates at the base of these plants, harboring many small animals including numerous worms, small snails and a variety of crustaceans. Large numbers of amphipods ranging in size from a quarter to a half inch occur here, as do many of small, usually juvenile, specimens of crabs including *Pugettia producta* (Figure 6-41) and *Cancer antennarius* (Figure 7-8). Interspersed among the algal turf are occasional large invertebrate predators. The Pacific sea star, *Pisaster ochraceus* (Color Plate 14b), is on hand in low numbers to feed on *Tegula* spp. The leather star, *Dermasterias imbricata* (Figure 6-22), is found here feeding on the aggregating anemones.

Surfgrass Flats

The red of the algal turf is occasionally broken up by the bright green of the surfgrass *Phyllospadix* (Color Plate 3b). Unlike the algae, which attach to bare rock with their holdfasts, surfgrass is a flowering plant that must eventually sink its roots into soft sediments to flourish. It is principally through the root system that surfgrass can take up nutrients necessary for continued growth, differing from the algae that absorb nutrients directly from the water across their entire surface. The root mass of the surfgrass traps suspended sediments and provides a habitat for many small animals, including peanut worms or sipunculids (Figure 7-2), polychaete worms of various types, and

Figure 7-2. Sipunculid worms. The larger specimen is four inches long and its introvert is fully extended, showing the feeding tentacles that surround the mouth.

small crustaceans. As sediments accumulate, the surfgrass sends out rhizomes trapping still more sediment and allowing it to increase its coverage.

On the individual blades of the surfgrass a red alga, *Smithora naiadum,* grows epiphytically (i.e., on the plant). In summer *Smithora* grows so densely that it may be difficult to see the green blades of surfgrass at all (Color Plate 3b). Also during the summer, a group of seasonal algae appears. The salt sac, *Halosaccion glandiforme (*Figure 6-12) and the large, fleshy red algae like *Iridaea* spp. and *Gigartina* spp. (Color Plate 3d) are particularly abundant.

Surfgrass Beds

The lower intertidal zone remains submerged much longer than the middle intertidal, and surfgrass can

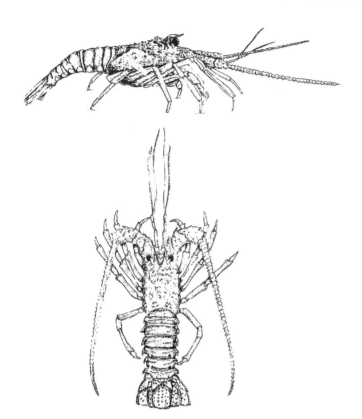

Figure 7-3. The California spiny lobster, Panulirus interruptus, *occasionally stranded in the low intertidal zone.*

Octopus. Although they probably spend low tide under or among adjacent rocks, octopuses are occasionally seen during low tide seeking the refuge of the surfgrass bed, especially if it is still partially covered by water. Octopuses are such masters of camouflage that they are often overlooked, and thus are considered more rare than they actually are. My best advice to beachcombers is not to set out purposely to find an octopus. Instead keep a ready eye when turning rocks or searching through algae and surfgrass. Octopuses are not only excellent mimics of their backgrounds, but also are very quick. When discovered, admire them from a distance, because some can give a good nip with their beaks and they do have a painful venom.

In northern and central California most of the octopuses seen are the small (up to 18-inch-long arms), red octopus, *Octopus rubescens.* Occasionally, small individuals of the much larger species (over 40 pounds and 10-foot arm spread in Pacific Northwest specimens), *O. dolfeini,* are seen. South of San Simeon in San Luis Obispo County, two very similar

grow quite lush here as a result. Extensive surfgrass beds that cover not only the solid rock substrate, but also small boulders and sandy areas as well are common sights in many rocky intertidal areas. Surfgrass is a favorite low-tide resting spot of large rock crabs, *Cancer* spp. (Figures 7-8 — 7-10), and often mated pairs can be discovered in the spring. Hermit crabs, *Pagurus* spp. (Figure 6-24), and kelp crabs, *Pugettia producta* (Figure 6-41), are also common. In southern California it is not unusual to find an occasional spiny lobster, *Panulirus interruptus* (Figure 7-3) under the surfgrass canopy. The lobsters are more common offshore, but are nocturnal, omnivorous foragers that sometimes get trapped in the low intertidal zone during very low tides. The surfgrass bed offers them a safe haven until the tide returns. The spiny lobster can get quite large offshore, but the average individual a beachcomber might find will probably be six to eight inches long. Don't underestimate them however! Even without claws they can cause nasty cuts and abrasions with their many sharp spines on their tail and antennae.

Figure 7-4. The common, small intertidal octopus of southern California, Octopus bimaculoides.

species are found in the intertidal zone, *Octopus bimaculatus* and *Octopus bimaculoides*. Both species sport two dark eye-spots ringed with blue (Figure 7-4). Octopus species are difficult to identify because of their ability to vary the color and the texture of their skin to match almost any background. They move about during high tide in search of prey that include crabs, shrimp, snails, and small fish. Octopus feed by trapping prey in the web of their arms and biting them with their bird-like beaks and injecting a nerve-deadening toxin. They will also feed on shelled mollusks by boring with their radula. In turn, octopuses are preyed upon by bottom-feeding fishes.

Nudibranchs. The surfgrass beds and the standing pools of kelp like oar-weed, *Laminaria* spp. (Figure 4-29) and other species, left by the lowest of the low tides, often harbor several nudibranch species among the vegetation. Nudibranch snails (sea slugs) apply the opposite color strategy compared to *Octopus*. These animals sport bright colors, often in elaborate patterns, supposedly to warn potential predators that they are unpalatable prey. A word about nudibranchs. On a successful day a trained observer may only discover eight or so species, so don't be too discouraged if you only see one or two. Nudibranchs are somewhat enigmatic to beachcombers. Their appearance (and disappearance) is unpredictable, so be advised to enjoy them when you can. I will mention only a few of the most common species. See more detailed references for coverage of all of California's nudibranchs [1, 2].

One of the most common nudibranchs in California is *Hermissenda crassicornis* (Color Plate 10b). It occurs on wharf pilings, in tidepools, and on mud flats. *Hermissenda* is beautifully and variably colored, but always has areas of orange on its back outlined with bright blue lines. The dorsal protuberances, called cerata, are found in rows behind the stalked sensory tentacles. *Hermissenda* feeds on hydroids but also will eat small anemones, bryozoans, worms, crustaceans, carrion, and even other *Hermissenda*. This shell-less snail can reach over three inches in length, but is usually smaller.

Nudibranchs like *Hermissenda* that have dorsal cerata are known as aeolid nudibranchs. Aeolid is a reference to the god of the wind, and it is refers to the cerata looking like leaves blowing in the wind. A second aeolid species that is seen in California is the pugnacious aeolid, *Phidiana pugnax* (Color Plate 10c). *Phidiana* has a translucent body bordered by a white foot. The cerata have a dark center banded with rose pink and topped with white or gold. It reaches two inches in length. The pugnacious aeolid feeds on cnidarians, but will eat almost anything else as well. When placed in a container with another nudibranch it immediately attacks, a behavior that gives it its common and specific names.

Another striking aeolid nudibranch occasionally encountered in southern and central California is the Spanish shawl nudibranch, *Flabellinopsis iodinea* (Color Plate 10a). This sea slug sports bright orange-red cerata contrasted against a vibrant purple body and can reach a size of three and half inches. It feeds on hydroids and compound tunicates. When disturbed, the Spanish shawl nudibranch is known to swim away by rapidly flexing its body sideways in a u-shape, first to one side and then the other.

A final aeolid nudibranch deserves a special note here. This is the shag-rug nudibranch, *Aeolidia papillosa* (Color Plate 11a and b). This species occurs sporadically all along the California coast and is a specialist on sea anemones. How much of a specialist is it? Take a look at the color illustrations. If you glanced at it in passing, you may have thought it was a sea anemone. Indeed, if this animal is in the middle of a clone of *Anthopleura elegantissima* (Color plates 4c and 4d), it is very hard to spot. The "shaggy" cerata on its back look just like a mass of anemone tentacles. The average animal seen is from one and half to three inches long, although four-inch specimens occur. The basic color is white to brown, although it can take on the color of the anemones it's eating. Compare the two individuals illustrated. I suspect the reddish specimen had been eating sea anemones with red pigment, like *Epiactis prolifera* (Color Plate 5b), which were abundant where this nudibranch was found.

More conspicuous, because of its size and bright color, is the sea lemon, *Anisodoris nobilis* (Color Plate 12a), which is bright yellow with black spots. This species can reach a size of over two and half inches in the intertidal zone, and much larger (up to six inches) specimens occur in the subtidal zone. Sea lemon nudibranchs feed on sponges. It is common north of Point Conception.

Another nudibranch commonly seen is the yellow-edged cadlina, *Cadlina luteomarginata* (Color Plate 10d). It has a pale-white color, edged and dotted in lemon yellow. Like the sea lemons, this species feeds

Figure 7-5. Hopkins rose nudibranch, Hopkinsia rosacea. *This inch-and-a-half-long sea slug matches its prey, the rosy bryozoan, in color.*

on sponges. It reaches a length of two inches and is able to transfer the sponge's spicules (glass skeletal elements) into its own skin for defense.

The ring-spotted dorid, *Diaulula sandiegensis* (Color Plate 12b), is named after one of my favorite California cities, San Diego. A common, sponge-feeding nudibranch found all along the California coast, *Diaulula* is unmistakable with its dark rings on a gray to light brown background. Beachcombers will usually see small specimens one to three and half inches in length, although animals five inches long have been reported.

A nudibranch frequently seen all along California that goes by the impressive common name "the white-spotted sea goddess" is *Doriopsilla albopunctata* (Color Plate 12c). This snail can reach almost three inches in length, but is usually smaller. It is yellow-orange to brown and has small protuberances on its back, each with a small white dot in the center, and

Figure 7-6. A three-inch-diameter colony of the rosy bryozoan, Eurystomella bilabiata. *Note the basket-weave texture that is typical of bryozoans.*

a ring of white gills towards the rear of its back. It also is a sponge feeder.

An unforgettable sight is the rose-pink Hopkins rose nudibranch, *Hopkinsia rosacea* (Figure 7-5), with its tall papillae gently swaying as it moves across the bottom. Hopkins rose feeds on a rose-colored animal called a bryozoan, *Eurystomella bilabiata* (Figure 7-6), which forms encrusting colonies on rocks. Bryozoans are found in most of the low intertidal habitats and can be recognized by the tiny reticulate (basket-weave like) pattern on their surface.

The sea-clown nudibranch, *Triopha catalinae* (Color Plate 12d), also feeds on bryozoans. The sea clown is usually one inch long, but can reach six inches, and is pale-white to yellow with deep orange spots on the body, gills, and all appendages. A close relative, the spotted triopha, *Triopha maculata* (Color Plate 12d), is often seen intertidally in its bright orange-red, juvenile form. It is usually less than a half inch in length. Larger individuals are generally subtidal in distribution.

BOULDER FIELDS

The Under-Rock Habitat

In low-lying regions of the middle and low rocky intertidal, small boulders will accumulate, forming boulder fields (Figure 7-7). In open, exposed rocky intertidal habitats small boulders are frequently moved around and broken up, or swept off the reef platform entirely by wave action. Therefore, boulder fields are more characteristic of rocky intertidal areas that are semi-protected from direct wave action. Here, rocks of dinner plate size and larger stand a good chance of staying in place. Thus, the under-rock habitat is available as a relatively protected, stable environment that supports a very rich and varied assemblage of plants and animals.

Look for rocks that have well-developed marine plant growth. This is an indication that they have remained in place for sometime, and will constitute fruitful searching. The common seaweed species will include the flat-bladed *Gigartina* spp. and *Iridaea* spp. growing in profusion (Color Plate 3d).

Under-Rock Crabs. The under-rock habitat is prime crab-viewing territory for the beachcomber. Prominent are the purple shore crab, *Hemigrapsus nudus* (Figure 6-55), and the large, red rock crabs, *Cancer antennarius* and *Cancer productus*; in southern California, the yellow rock crab, *Cancer anthonyi* is seen.

Cancer antennarius (Figure 7-8) is the most abundant *Cancer* crab found in the intertidal zone in southern and central California. This rock crab is brick red on its upper surface and is distinguished from the similarly colored *Cancer productus* by the presence of red speckling on its lower surface. *Can-*

Figure 7-7. A middle intertidal zone boulder field.

cer antennarius feeds on both live and dead animals. It preys on hermit crabs by chipping away their shell with its large claws until the crab is exposed. It also feeds on snails, other crabs, and carrion. Males grow larger than females and reach five inches across the carapace (back). Females carrying eggs can be found in the winter months and during early spring. Beach-combers should be very careful with all of these large *Cancer* crabs as they can deliver a very painful pinch.

Cancer productus (Figure7-9) is found only occasionally in the intertidal zone, and appears to migrate offshore during the winter. *Cancer productus* is brick red in color with black-tipped claws similar to the closely related *Cancer antennarius*. However *C. productus* does not have red speckles on its underside like *C. antennarius*. Young *C. productus* are frequently light colored with stripped or mottled color patterns like the animal shown in Figure 7-9. Occasionally, a gray to dark brown color variation of this crab is seen and looks very similar to *Cancer gracilis* (Figure 4-21). The two can be distinguished by the color of the tips of their claws: *C. gracilis* has white-tipped claws, while *C. productus'* claws are tipped with black. Male *Cancer productus* can reach a carapace width of seven inches, females are smaller. This crab feeds on barnacles, small crabs, and dead fish among a variety of other prey. Although it occurs all along the California coast, it is more abundant north of Point Conception.

Figure 7-9. A juvenile rock crab, Cancer productus. *Note the prominent stripping on the three-inch-wide carapace of this male specimen.*

The yellow crab, *Cancer anthonyi* (Figure 7-10), can be found under rocks in the intertidal zone from San Pedro (Los Angeles County) south. It occurs in the subtidal zone as far north as Humboldt County. *C. anthonyi* reaches a size of six inches across the carapace, is pale yellow in color, and lacks any speckling on its underside. The claws are tipped in black. All three of these large *Cancer* crabs are trapped subtidally by crab fishermen using baited hoop nets. A

Figure 7-8. Large male rock crab, Cancer antennarius. *The large size of the black-tipped claws are diagnostic of a sexually mature male.*

fishing license is required to catch them, and they are subject to California Fish and Game season and size regulations.

The under-rock habitat is home to several genera of pebble crabs. The smaller (one inch across the carapace) crabs of the genus *Lophopanopeus* (Figure 6-30) are sometimes abundant, especially if the rocks rest on a bottom of mixed sediment and pebbles. There are several species of *Lophopanopeus* along the California

coast, and they are difficult to distinguish. Consult more extensive sources [3, 4] to sort them out.

From Monterey south, two larger pebble crabs can be found. The first is the lumpy pebble crab, *Paraxanthus taylori* (Figure 7-11) which can reach one and a half inches in carapace width. The lumpy pebble crab is dark red above, and has dark-brown tipped claws with conspicuous knobby bumps. It is particularly abundant in the San Diego area.

The large pebble crab *Cycloxanthops novemdentatus* (Figure 7-12), is also common in the San Diego area, and less abundant in the northern part of its range. This crab can reach almost four inches across the carapace in the males, and is brown tinged with red or purple; the claws are black. Like all the pebble crabs, the large pebble crab feeds mainly on coralline algae and other red algae, although some animal material is also eaten. Large pebble crabs have been observed breaking open and eating purple sea urchins, *Strongylocentrotus purpuratus* (Color Plate 15a). Also characteristic of the pebble crab clan, the large pebble crab will often "play dead" when handled, remaining stiff and motionless for some minutes.

Hordes of the flattened porcelain crab, *Petrolisthes* spp (Figure 6-65), occur under rocks in some places. Also found here will be large (three inches across the carapace) kelp or shield back crabs, *Pugettia producta* (Figure 6-41). From Santa Barbara south, the southern or globose kelp crab, *Taliepus nuttallii* (Fig-

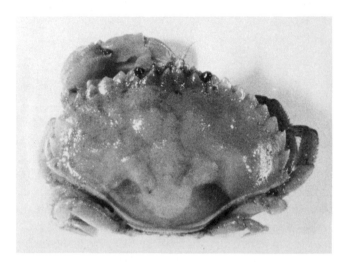

Figure 7-10. The yellow cancer crab, Cancer anthonyi. *This crab is found in the intertidal zone only in southern California.*

Figure 7-11. Lumpy pebble crab, Paraxanthus taylori, *one inch across the carapace. Bumps on claws give it its common name.*

ure 7-13), may also show up. *Taliepus* is dark brown or purple in color and can reach three and a half inches in carapace width. It is distinguished from *Pugettia* by its stout legs and rounded body. Both kelp crabs have some algae in their diets, but the globose kelp crab prefers brown algae like *Egregia* and *Macrocystis,* while the shield back crab adds animal tissue to its diet as it grows older.

In southern California, young decorator crabs, *Loxorhynchus crispatus*, occasionally turn up (Figure 7-14). These young crabs are relatively clean compared to subtidal adults which are truly decorated. These crabs adorn their bodies with bits and pieces of algae and encrusting invertebrates so that they look like part of the bottom.

Occasionally, while turning rocks in the low intertidal zone, the beachcomber is treated to an unusual animal known as the umbrella crab, *Crytolithodes sitchensis* (Color Plate 13d). The umbrella crab is also known as the turtle crab, both common names referring to the fact that the carapace extends out on the sides to hide the legs. This crab reaches almost three inches in carapace width, although usually smaller individuals are seen. It varies in color from the bright orange animal illustrated here to individuals with highly mottled color patterns. Very little is known about the biology of this strange-looking crab.

Under-Rock Serpent/Brittle Stars. If there is any sediment beneath the rock, the beachcomber should look for sipunculid worms, shallow-burrowed polychaete worms, and brittle or serpent stars. In addition to the three species mentioned in Chapter 6, several other species can be found. "Serpent" and "brittle" stars are closely related and belong to the echinoderm subclass Ophiuroidea (ophiuroids). The two common names roughly distinguish between those species that readily shed their arms when handled (thus "brittle" stars) from those that don't. The common name "serpent" star refers to the jerky, snake-like movement of the ophiuroids' jointed arms.

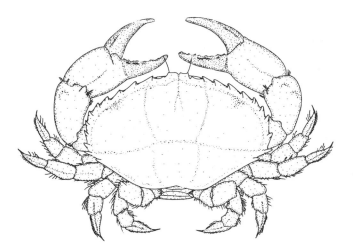

Figure 7-12. *The large pebble crab,* Cycloxanthops novemdentatus, *a common intertidal crab in southern California, can reach four inches in width.*

Figure 7-13. *The globose kelp crab,* Taliepus nuttallii, *can reach three and a half inches in width. It is also known as the southern kelp crab.*

Figure 7-14. *A small specimen of the masking crab,* Loxorhynchus crispatus. *Larger, heavily "decorated" individuals are found in the subtidal zone.*

Figure 7-15. The hearty Esmark's serpent star, Ophioplocus esmarki. *The disk is three quarters of an inch in diameter.*

Figure 7-16. The large, blunt spines of the serpent star, Ophiopteris papillosa, *are unmistakable. This specimen is five inches in diameter.*

One of my favorite ophiuroids is Esmark's serpent star, *Ophioplocus esmarki* (Figure 7-15), which is found in the low intertidal from Tomales Bay in Sonoma County south. This rather stout star reaches over an inch in disk diameter, and the arms extend another two to three inches. It is solid brown to red brown in color and feeds on animal tissue.

Another relatively common species is the spiny brittle star, *Ophiothrix spiculata* (Color Plate 14d), which occurs from Moss Beach (San Mateo County) to Peru. The central disk is about a half inch in diameter and covered with short spines that give it a fuzzy appearance. The length of the arms is five to eight times the diameter of the disk, and they have many long, pointed spines. The spines themselves are secondarily adorned with spinlets. This is truly one spiny character! The color varies, with the most common color being green with orange bands around the arms.

The spiny brittle star is primarily a filter feeder, trapping suspended material with mucous strands held by its upwards stretched arms. It can also feed on bottom detritus and capture live prey.

A final under-rock serpent star of the low intertidal zone is the flat-spinned *Ophiopteris papillosa* (Figure 7-16). Like the spiny brittle star, *Ophiopteris* is a versatile feeder, employing mucus for suspension feeding as well as being an active carnivore. The disk is up to an inch an a half in diameter, with the arms three to four times the disk diameter in length. All the spines are thick and blunt with the spines towards the tips of the arms being quite short. This serpent star is a deep chocolate brown in color, and occurs all along the California coast.

Under-Rock Sea Stars. Several other echinoderms frequent the under-rock habitat. The sea bat, *Patiria miniata* (Figure 6-28), can be seen on the tops and sides of rocks, but only under the rocks will you find small sea bats. The small six-rayed sea star, *Leptasterias* sp. (Figure 6-27) is common here also as far south as the California Channel Islands. The blood star, *Henricia leviuscula* (Figure 7-17), shows up here under rocks. It is called the blood star because large specimens (seven inches in diameter) are orange-red in color, smaller (two to three inches in diameter) individuals are gray, tan, or purple, often banded with darker shades. The blood star feeds by trapping bacteria and other tiny particles in mucus on its arms, and transporting it to the mouth by way of ciliary tracts. It may also feed on sponges and bryozoans.

The Pacific sea star, *Pisaster ochraceus* (Color Plate 14b), can be common, and occasionally the giant pisaster, *Pisaster giganteus* (Color Plate 14c), occurs. When found in the intertidal zone, *Pisaster giganteus* rarely exceeds a diameter of ten inches, although individuals two feet in diameter occur subtidally. In the intertidal zone this species feeds mainly on mollusks, especially mussels and snails. Color varies in *P. giganteus*, but it can be distinguished by large, white, evenly spaced spines on its aboral (side opposite the mouth) surface that are surrounded by a zone of blue.

You may come across two multi-rayed sea stars in central and northern California. The first is the sunflower star, *Pycnopodia helianthoides* (Color Plate 14a), discussed in Chapter 6. Sunflower stars are more at home subtidally, but they do forage into the

Figure 7-17. The blood star, Henricia leviuscula. *Adults of this star are bright orange red in color.*

Figure 7-18. The multi-rayed sea star, Solaster stimpsoni. *Note the prominent color pattern radiating from the disk onto the arms.*

intertidal zone, and become stranded when the tide recedes, especially if it is a very low tide. In northern California, you can find Stimpson's sun star, *Solaster stimpsoni* (Figure 7-18), among low intertidal rocks. This sea star usually has ten arms and reaches a diameter of eight inches. Its color varies with red, orange, yellow, green, and blue individuals seen. However, it is readily recognized by a blue gray spot on its central disk that continues as a distinct stripe onto the arms. *Solaster* is a fairly specialized feeder, preferring sea cucumbers.

Sea Cucumbers. These creatures are different from all the other echinoderm groups. Instead of being star-shaped or round, they are elongated and worm-like. Also, they are very flexible, not rigid like a sea star or sea urchin. These qualities make sea cucumbers very successful dwellers of tight quarters, like the under-rock habitat. The most obvious sea cucumber you will find in southern California is the warty character known as *Parastichopus parvimensis* (Color Plate 15c). This animal can reach a length of at least ten inches and is chestnut brown in color. It has a series of blunt tube feet around its mouth that are used to pick up deposited material from the bottom sediment. The warty sea cucumber moves on a "sole" of tube feet, and the rest of the body is adorned with stout warts tipped with black. Other smaller sea cucumbers, which are filter feeders, occur in the rocky intertidal zone. One particularly colorful animal is the small (less than one inch long), red *Lissothuria nutriens* (Color Plate 15d). This sea cucumber is found on the sides of rocks and among the holdfasts of algae and the roots of surfgrass.

A relatively common sea cucumber under rocks in northern California is the creamy white *Eupentacta quinquesemita* (Figure 7-19). This animal is two to three inches long and all five rows of tube feet are still visible on its elongated body. *Eupentacta* is also a filter feeder, but only extends its branched feeding tentacles when submerged. You may also find this animal on floats and pilings, especially in Humboldt Bay [5].

Small (less than an inch in diameter) purple sea urchins, *Strongylocentrotus purpuratus* (Color Plate 15a), are found under the rocks, and frequently have green rather than purple spines. Small urchins are seldom seen among the larger individuals in other habitats, and it appears that they use the under-rock habitat along with the recesses of the mussel clump as nurseries.

Animals on the Undersides of Rocks. Attached to the bottom of the rocks will almost certainly be the agile, proliferating sea anemone, *Epiactis prolifera* (Color Plate 5b), in one or several of its variety of colors. You may also find the calcium carbonate tubes of two common, filter-feeding polychaetes worms. The small, spiral tubes are made by *Spirorbis* spp. (Figure 6-39) already discussed. The larger (three inches), coiled or meandering tubes are made by the fan-worm, *Serpula vermicularis* (Color Plate 7d). *Serpula*'s tube is often partially concealed in a crack or

Figure 7-19. The sea cucumber, Eupentacta quinquesemita, *with its anterior feeding tentacles extended.*

Figure 7-20. The highly ornamented, three-inch-long shell of the tube snail, Serpulorbis squamigerus. *The smaller, spiral tubes belong to the polychaete Spirorbis.*

abandoned burrow in the rock. When submerged, *Serpula* holds a bright-colored fan of red or orange filter-feeding tentacles into the water. Cilia on the tentacles trap plankton and other detritus and carry it to the mouth. One of the tentacles is modified into a knob-like stopper or operculum, which *Serpula* pulls in last to seal off its tube.

Another animal that dwells in a meandering, calcium carbonate tube in this habitat is the scaly tube snail, *Serpulorbis squamigerus.* These snails were introduced in Chapter 5, because they are found on rocks in bays and estuaries. They can also be very abundant on the sides and under rocks, and on the walls of surge channels. *Serpulorbis* is found intertidally from Monterey to Baja California. Solitary individuals (Figure 7-20) are often found in the northern part of its range, while in southern California, great masses of twisted, intertwined shells can be seen (Figure 5-47). These snails are filter feeders and secrete long strands of mucus that they use to trap suspended material.

The tube-like shell of the tube snail can be distinguished from the tube of the polychaete *Serpula vermicularis* by the fact that it is larger in diameter (half an inch versus a quarter inch), consists of three layers instead of just one, and is ornamented with fine, scaly striations and concentric rings, while the worm's tube is smooth. The tube worm closes off its tube with a knob-like stopper. The tube snail closes off its shell

Figure 7-21. Shells of the scallop Hinnites giganteus. *The lower shell is from a free-living juvenile, upper shell is from an attached adult scallop.*

with the bottom of its foot, which is conspicuous because of its dark center surrounded by a white ring that is surrounded by an orange-red ring.

Also seen attached to the under-rock surface are a variety of bivalve mollusks. By far the largest (up to six inches in diameter) is the rock scallop, *Hinnites giganteus*. *Hinnites* lives freely as a yellow or orange-shelled juvenile up to a size of almost two inches, and has a typical scallop appearance (Figure 7-21). At a size between one to two inches, the small scallop attaches by cementing its right valve to the rock. Once attached, the once beautifully symmetrical

bivalve grows along the contours of the rocky surface and assumes a variety of shapes depending on the attachment site (Figure 7-21). The detached left valve is frequently found by beachcombers, and the remnant of the juvenile shell can still be seen near the hinge, along with a deep purple stain on the inside of the shell that identifies it as a *Hinnites'* shell. In life the animal is quite spectacular, with bright orange flesh and silver blue eyespots all along its outer margin.

Another flattened bivalve mollusk, the jingle shell, *Pododesmus cepio* (Figure 7-22), occurs under larger

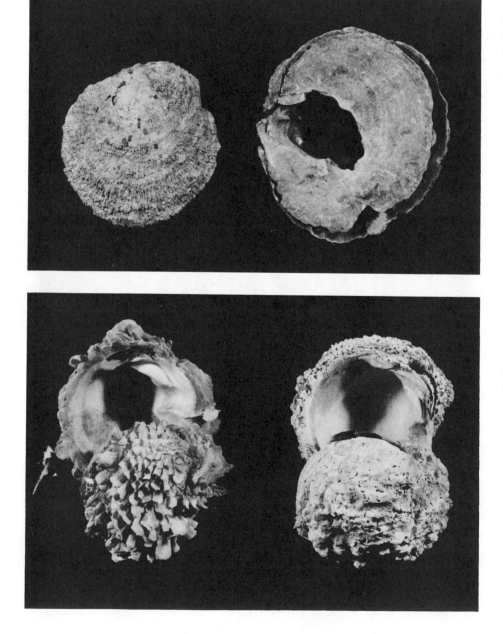

Figure 7-22. Views of the upper (on left) and lower valves of the rock jingle, Pododesmus cepio. *Note the deep notch in the intact shell on right.*

Figure 7-23. Two rock chamas, Chama arcana, *one and a half inches in diameter. Note the bottom (left) valve is deeply dished while the top valve is flat.*

The Rocky Intertidal Environment: Middle and Low Zones 155

rocks. It is also occasionally found attached to the shell of red abalone, *Haliotis rufescens*. This distant relative of the scallop attaches to the rock with a large calcified attachment (byssus) that projects through a notch in the flat, lower (right) valve. The right valve is thin and conforms to the rock's surface, while the upper (left) valve is concave and can reach a diameter of three inches. Jingle shells are named for the detached, thin left valve that bounces and "jingles" in the surf.

Two other attached bivalves should be noted here. These are known as jewel boxes or chamas. The first is the right-hand chama, *Chama arcana* (Figure 7-23), and the second is the reversed chama, *Pseudochama exogyra*. The jewel boxes are so named for their thick, deep, lower valves that are topped by the much thinner, flatter top valves like the lid on a box. The valves of most specimens are decorated with frilly or spiny projections, but can also be fouled with marine growth. Both these mollusks can reach two inches in diameter. They regularly co-occur all along the California coast. They are distinguished by looking down onto the top valve. *Chama arcana* (Figure 7-23) attaches by its left valve, and the top valve (the right valve) coils to the right. *Pseudochama exogyra* attaches by its right valve and the top valve (left valve) coils to the left. Like the scallop and jingle shell previously described, the chamas are regularly found on a variety of hard substrates including breakwater rocks, harbor pilings, and the undersides of harbor floats. From Monterey south, the chamas regularly co-occur with the tube snail *Serpulorbis*.

Chitons. Various chitons are found on the rocks. The shells of these mollusks are divided into eight articulating plates that allow the chitons to conform and cling to irregular rock surfaces. The obvious mossy chiton, *Mopalia muscosa* (Figure 6-20), also occurs here along with a close relative, *Mopalia lignosa* (Color Plate 8a). *M. lignosa* is fairly common on the sides and under large boulders from Alaska to Point Conception. Its valves are beautifully colored with greens and browns.

From Mendocino County south, the elongated *Stenoplax heathiana* (Color Plate 8b) is a common sight, especially if the rock is partially buried in sand or fine gravel. *Stenoplax* reaches a length of four inches and is twice as long as wide. It is pale gray to brown cream in color and often spotted with light or

Figure 7-24. The gumboot chiton, Cryptochiton stelleri. *The world's largest chiton.*

dark greenish gray. The girdle is buff to brown in color. This chiton remains hidden under the rock by day and emerges at night to feed on drift algae that lodges at the base of the rocks. South of Santa Barbara a close relative with similar nocturnal habits, the conspicuous chiton, *Stenoplax conspicua*, can also be seen.

Occasionally the beachcomber will find what appears to be a small football or misplaced meatloaf among the rocks in the low intertidal zone. Congratulations, you just found the world's largest chiton, *Cryptochiton stelleri* (Figure 7-24). Commonly called the gumboot chiton, this animal can reach a length of 14 inches, and occurs from Alaska to as far south as California's Channel Islands. It is dark brown to brick red in color, and the stiff mantle completely covers the chiton's eight valves (shells). The gumboot feeds on fleshy red algae and certain coralline algae, and occasionally on brown and green algae. Unfortunately, the gumboot does not attach as firmly to the rocky substrate as other chitons, and frequently is washed out of the lower intertidal and stranded on the high beach during storms.

If there are any encrusting coralline algae on the rocks, you may discover one of California's true marine jewels, the strikingly colored, lined chiton, *Tonicella lineata* (Color Plate 8c). This small (up to two inches long), beautiful chiton is common in low intertidal habitats that support its food, coralline algae. *Tonicella*'s shells vary in color from purple to light red marked with zigzag lines of alternating colors in combinations of dark and light red, dark and light blue and red, or whitish and red. The stiff, fleshy girdle surrounding the shells is often alternately banded as well. The main predators of the lined chiton are the sea stars *Leptasterias* and *Pisaster ochraceus* (Color Plate 14b).

California has one of the richest chiton faunas in the world, and the list is too long to cover in this guide. I'll leave you with another of my favorites, *Lepidozona mertensii* (Color Plate 8d), and refer you to more complete references for the rest [3, 4]. *Lepidozona* reaches a length of two inches, and is orange-

red to brick-red to crimson in color. It is sometimes marked with white blotches as is the handsome animal illustrated in Plate 8d.

Under-Rock Snails. Several shelled gastropods occur among the rocks of the middle and low intertidal zones. They are too numerous to include in this guide, so I'll stick with the largest, most conspicuous animals. The largest snails by far are the abalone. The black abalone, *Haliotis cracherodii* (Figure 6-54), can

Figure 7-25. Five-inch red abalone, Haliotis rufescens, *showing prominent podial tentacles. Shell is encrusted with acorn barnacles and a bryozoan colony.*

Figure 7-26. A six-inch-diameter shell of the green abalone, Haliotis fulgens. *Note the round, slightly raised openings in the shell.*

be found here under rocks. In central and northern California small (one to three inches long) red abalone, *Haliotis rufescens* (Figure 7-25), use the boulder fields as a nursery, with an occasional large (up to 12 inches!) animal also appearing. However, because of human fishing pressure, the large red abalone are usually found only in the subtidal zone. South of Point Conception red abalone are exclusively subtidal. The red abalone shell is usually brick red, and the surface uneven with three to four open holes. The holes are oval and slightly raised above the surface of the shell. The color of the shell is influenced by diet, and it is frequently fouled and eroded by marine growth. The red abalone is a favorite food of the California sea otter. Sea stars, fishes, and *Cancer* crabs are other predators.

Red abalone are replaced somewhat in the southern California intertidal zone by the green abalone, *Haliotis fulgens* (Figure 7-26). The green abalone's shell reaches up to ten inches in length. The exterior of the shell is olive green to reddish brown in color, and often with fine radiating ribs. The shell has from five to seven open holes, which are round and slightly raised above the surface of the shell. The interior of the shell is strongly iridescent with areas of dark green and blue, with pink and purple highlights. Like the red abalone, the green abalone feeds primarily on drifting algae, preferably kelp. Its chief predator is the octopus.

Another group of large snails found among low intertidal boulders are closely related to the abalone. These are the keyhole limpets. Keyhole limpets possess a respiratory opening at the apex of their shell that gives it a "keyhole" appearance. Water that comes in over the limpet's head and through its gills passes out this opening that serves the same function as the holes in an abalone shell. Figure 7-27 shows the shells of California's three common keyhole limpets.

Diadora aspera (Figure 7-28), the rough keyhole limpet, is common in northern and central California, and mainly subtidal in the south. The rough keyhole's shell can be over two and a half inches long and gray in color with coarse black and white radiating ribs. The opening is circular. It feeds on algae and encrusting animals such as bryozoans. Rough keyhole limpets are the prey of sea stars and have a well-developed escape response that often foils their predators. When touched by the tube feet of a sea star predator, they rapidly expand their fleshy mantle up

Figure 7-27. Shells of three California keyhole limpets. Clockwise from the top: Megathura crenulata, Diadora aspera, *and* Fissurella volcano.

over their shell, leaving no firm surface for the sea star to grab.

The volcano keyhole limpet, *Fissurella volcano* (Figure 7-29), is locally abundant all along the California coast under rocks in the middle and low intertidal zones. The conical shell of this snail is up to an inch and a half long and the opening is elongate and narrow. The shell exterior is pink with black or reddish brown rays. The shell interior is greenish. The volcano limpet's chief predator is the sea star *Pisaster ochraceus* to which it shows a running escape response.

Next to the abalones, the most impressive intertidal snail is the giant keyhole limpet, *Megathura crenulata* (Color Plate 9a). The giant keyhole's shell is thick and up to five inches long with fine radial and concentric cords that produce a finely beaded texture. It is light gray or tan in color and relatively flat, with a prominent oval opening just forward of center. However, the shell rides on the snail's back like a saddle rather then providing it with a sheltering retreat. The snail can reach ten inches in length and its fleshy mantle usually completely covers the shell. The mantle itself may be black, yellowish brown, or a mottled gray. The foot is orange or bright yellow. Giant keyhole limpets eat algae and encrusting marine inverte-

Figure 7-28. Two-inch-long rough keyhole limpet, Diadora aspera. *Note the strong pattern of radiating ribs.*

Figure 7-29. Top view of a one-and-a-half-inch keyhole limpet, Fissurella volcano. *Note the elongated shell opening and the mantle tissue below.*

brates. They are seen in the intertidal more frequently in southern California among low intertidal rocks, and are subtidal in the northern part of their range that extends to Monterey.

South of Point Conception a pair of large, active, plant-eating snails can be found, especially if there is a kelp bed nearby. The largest of these is the wavy turban snail, *Astraea undosa* (Figure 7-30). The shell is up to four inches high and often partially or com-

pletely covered with marine growth. The shell is conical or top-shaped, and has a series of strong spiral and vertical wavy ridges. The aperture (opening) of the shell is pearly, and closed off with a distinctive operculum in living snails. The operculum (Figure 7-30) is calcified and teardrop-shaped, with three prominent ridges on its outer surface. You may more often find the empty shell or detached operculum on the beach than the animal itself.

Figure 7-30. Top and bottom view of the wavy top, Astraea undosa. *Specimen on left is one and a half inches across the bottom. Note the teardrop-shaped operculum.*

The second, southern plant eater is Norris's top snail, *Norrisia norrisi* (Figure 7-31). The beach-comber can't miss this snail because its bright, red-orange foot contrasts vividly against its drab, brown algal food. Norris's snail occurs in the low intertidal zone among algae-covered rocks, and is common off-shore in kelp beds. The shell is deep chestnut brown, and up to two inches in diameter and an inch and three quarters high.

Among the coralline algae-encrusted rocks, the dunce cap limpet, *Acmaea mitra* (Figure 7-32), can be found. This limpet's steep, conical shell is white, although often covered with the same pink encrusting coralline algae on which it feeds. The shell can reach an inch and half long and over an inch high. This snail's tall shell appears top heavy and to be a liability in a wave swept habitat. However, its foot is able to produce a remarkably strong attachment to the rocks and it stays put!

The white slipper limpet, *Crepidula nummaria* (Figure 7-33), is more obvious in death than when alive because its empty shell is frequently found on the beach or in a tidepool. This snail is related to *Crepidula adunca* that rides around on *Tegula* shells, and like it is a filter feeder. It lives under rocks or in the empty holes of burrowing animals or in empty shells. The white slipper limpet's shell is white and up to an inch and half long. The shape of the shell varies as it conforms to the contours of the animal's living

Figure 7-31. Norris' top snail, Norrisia norrisi, *common in kelp beds and sometimes found in the low intertidal zone in southern California.*

space. The shell's exterior is often covered with a thick thatched material known as periostracum. The interior of the flat shell sports a prominent shelf, which gives the shell its slipper-like appearance. This shelf supports the limpet's large gills that are used in filter feeding.

The ornately shelled, carnivorous snail known as the leafy hornmouth, *Ceratostoma foliatum* (Figure 7-34), is found in central and northern California. This snail receives its common name from the elaborate "leafy" flanges on its shell and the sharp tooth that protrudes from the base of its aperture (shell opening). These flanges are more pronounced in subtidal individuals, and their function is poorly understood. Leafy hornmouths can reach a length of four inches, although intertidal specimens are usually two inches or smaller. They can be found among the rocks of boulder fields, and on the sides of rocky outcroppings and surge channels. The chief prey items of this carnivorous snail are acorn barnacles (*Balanus* spp.) and bivalves. South of Point Conception the closely related Nuttall's hornmouth, *Ceratostoma nuttallii*, is found. Nuttall's hornmouth is very similar in diet and appearance to the leafy hornmouth, although the flanges are not as elaborate and the shell only reaches two and a half inches in length. The shell is white to brown with darker shells often banded in white.

A southern California snail closely related to the hornmouth is the striking three-winged murex,

Figure 7-32 (Top). The dunce cap limpet, Acmaea mitra. *The shell is covered with the snail's food, encrusting coralline algae.*

Figure 7-33 (Middle). Shells of the slipper limpet, Crepidula nummaria, *often found empty on the beach.*

Figure 7-34 (Bottom). The leafy hornmouth, Ceratostoma foliatum, *of central and northern California.*

Figure 7-35. The showy shell of the three winged murex, Pteropurpura trialata, *from southern California. Shell on left is one and a half inch es high.*

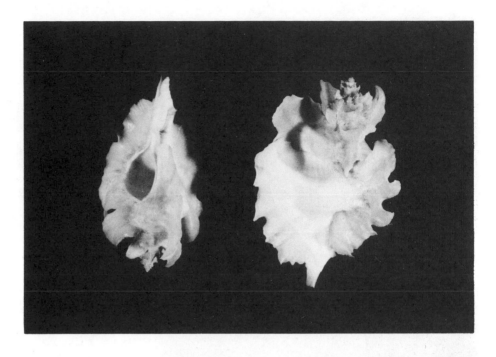

Pteropurpura trialata (Figure 7-35). This snail can reach about three inches in length and its three flanges or "wings" can be quite frilly and ornate. The shell is white with brown between the wings, and it lacks the hornmouth's projecting tooth at the aperture. *Pteropurpura* is a carnivore that feeds on sessile gastropods like *Crepidula* and *Serpulorbis*.

While the three-winged murex is common among low intertidal rocks, a close relative, the festive murex, *Pteropurpura festiva* (Figure 7-36), occurs there only occasionally. This snail is more common on muddy rocky areas of bays. It reaches a height of two inches, and the edges of the "wings" of this murex are characteristically rolled backwards. This snail is found from Santa Barbara south to Baja California.

Three snails seen occasionally among the low intertidal rocks in southern California are the sole representatives of groups that are very abundant in tropical and sub-tropical faunas. These snails are the California cone snail, *Conus californicus* (Figure 7-37), the chestnut cowry, *Cypraea spadicea* (Figure 7-38), and Ida's miter, *Mitra idae* (Figure 7-39). All of these snails are seen occasionally in the low intertidal zone of southern California, and occur only subtidally in the northern limit of their range, Monterey.

The shell of the California cone is a pale imitation of the glorious shells of its tropical relatives. It is gray brown in color, conical with a long narrow aperture (Figure 7-37), and up to an inch and a half in length.

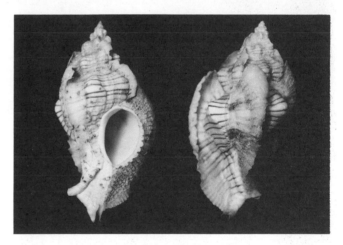

Figure 7-36. The festive murex, Pteropurpura festiva. *Note the rolled-back edge of the aperture characteristic of this species.*

As with all cone snails, the California cone subdues its prey by firing a sharp, harpoon-like tooth into its prey and injecting poison into the wound. The California cone feeds on polychaetes, snails, and bivalves. It is a nocturnal predator, so you will most likely only see this snail if it is discovered under a rock, or an empty shell is found on the beach or on a hermit crab.

In contrast to the homely cone, the chestnut cowry, *Cypraea spadicea*, is a beauty (Figure 7-38). The shell, which can reach almost three inches in length,

Figure 7-37. Shells of the California cone snail, Conus californicus. *Larger shell is one inch high.*

Figure 7-39. The tall dark shell of Ida's miter, Mitra idae, *contrasts vividly with its white body.*

Figure 7-38. Top and bottom views of the chestnut cowry, Cypraea spadicea. *Note the long slit-like aperture. Shells are two inches long.*

has a smooth, glossy appearance with the top being mostly brown and the edges gray or white. The bottom of the shell is white, and has a slit-like aperture (opening) bordered by denticles (tooth-like ridges). In life the snail has a bright orange mantle with dark spots, which it can expand to cover its shell completely. It is a spectacular sight! Chestnut cowries are

never abundant, but can be found in the low intertidal among rocks covered with seaweed. They are carnivores, feeding on anemones, sponges, and other animal tissue.

Mitra idae presents a striking contrast between its tall black shell and pale white body (Figure 7-39). The shell is actually dark brown, but is usually covered by a heavy black periostracum, a layer of the shell that normally wears off most mollusks. Ida's miter can reach over three inches in length, and is occasionally found in the low intertidal of southern California among rocky rubble. It is much more common in the shallow subtidal, especially under kelp beds. Ida's miter occurs from Crescent City in California's northernmost Del Norte County to Baja California.

Two other southern California snails deserve brief mention. Both these species can be locally abundant among the rocks in the low intertidal zone south of Santa Barbara. The first is a pretty animal known as Maxwell's gem murex, *Maxwellia gemma* (Figure 7-40). The shell is white or gray with brown or black spiral cords wrapping around it. The shell can reach almost two inches in length. This snail lives in the very low intertidal zone and is common on breakwaters near the entrance to bays. Its carnivorous habits are not well known.

The second southerner is Poulson's rock snail, *Roperia poulsoni* (Figure 7-41). This snail is found on rocks in tidepools, but is more common in bays than on the open coast. The shell can reach over two inches in length, although it is usually half this size. The striking shell is off-white in color, and has heavy,

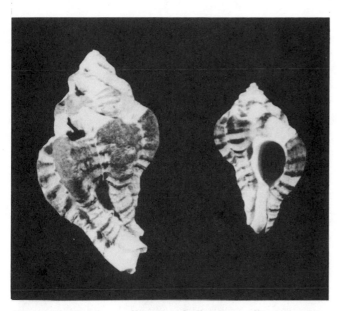

Figure 7-40. Maxwell's gem shell, Maxwellia gemma. *The specimen on the right is three quarters of an inch high.*

Figure 7-41. Poulson's rock snail, Roperia poulsoni. *Specimen on left is two inches in height. Note the fine brown striations around the shell.*

knobby, vertical ridges crossed with broad white spiral cords separated by fine brown spiral grooves. This snail is a "drill," which uses its radula to bore into the shells of its prey: mussels, barnacles, and snails.

The Rock Borers

The rocks and boulders found in intertidal boulder fields are derived from a variety of sources. Some material may come from the weathering of cliffs that abut the intertidal zone. Much rocky debris may be thrown onto the reef from the subtidal zone by wave action. However, most of the rocks are usually composed of pieces broken from the solid reef substrate itself, especially if the reef is made up of soft, sedimentary rock like sandstone or siltstone. If the rocks of the boulder field are soft, you might notice round holes in them. These are the burrows of the rock-boring bivalve mollusks. The burrows excavated by these unique clams appear round from the surface, and conical when viewed from the side (Figure 7-42). These clams riddle the soft, rocky reef in the mid-and lower-intertidal zones until it becomes unstable, and pieces break away under the pounding surf.

There are several species of bivalve borers on California's reefs, and the most abundant is the common piddock, *Penitella penita* (Figure 7-42). Occasionally, a detached piece of reef rock will contain a live specimen still in place in its burrow, but more often only the elongated, augur-shaped shells remain. Piddocks or pholads, as they are also called, burrow by mechanical action. The anterior edges of their shells are laid down as a series of stout ridges, which function as drill blades to cut into the soft reef rock. The clams anchor themselves by their foot with the anterior end of their shell pressed against the bottom of their burrow. They then rotate their shells around the foot by alternate contractions of the shell muscles, and a grinding, drilling action results. The worn shell ridges are replaced by the secretion of new shell. In this way the clam excavates a conical burrow into the rock, which increases in diameter as the clam grows. Once the clam becomes sexually mature the burrowing ceases. If enough clams have burrowed in proximity, the rock is severely weakened and subject to fracturing.

Another, less-common boring clam is the rough pea-pod borer, *Adula falcata* (Figure 7-43). Beachcombers will find the long, narrow, fragile shells of this clam in the same soft rocks with *Penitella*. The pea pod borer also uses its shell to excavate the rock, but it does not rotate the shell while it is boring, so it fits much more snugly in its burrow.

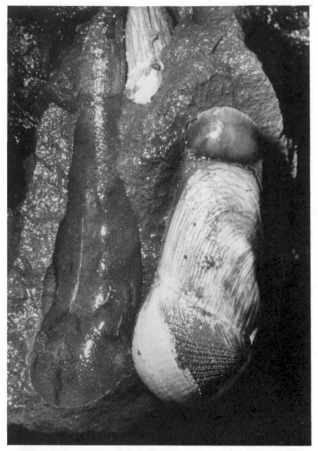

Figure 7-42. A two-and-a-half-inch-long rock piddock, Penitella penita, *shown next to a cut away view of the burrow it has excavated in the soft rock.*

Figure 7-43. The rough pea-pod boring clam, Adula falcata, *shown next to its partially-exposed burrow. This clam is five inches long.*

Figure 7-44. A lumpy porcelain crab, Pachycheles rudis, *three quarters of an inch across the carapace, found in the empty burrow of a boring clam.*

When the boring clams die, the now vacant burrows become the home of a variety of other animals. Sipunculids, polychaetes, snails, small sea stars, nestling clams, and encrusting animals like sponges and bryozoans all can show up. Probably the most surprising animal to find here is the lumpy porcelain crab, *Pachycheles rudis* (Figure 7-44). This small (one inch across the carapace) crab often occurs in mated pairs, and sometimes grows too large to escape its tapered abode. It is also common in mussel clumps and in the holdfasts of large seaweeds. Like the smooth porcelain crabs (Figure 6-65), these animals are filter-feeders, so they only need an open connection with the circulating water to feed successfully.

LOW INTERTIDAL POOLS AND SURGE CHANNELS

There are two remaining habitats that deserve mention, the low intertidal pool and surge channels. Surge channels technically are not just a low intertidal habitat, because they can extend all the way to the highest intertidal zone; however, they have their origin in the low intertidal and will be treated from that perspective.

Low Intertidal Pools

Low intertidal pools are unique habitats. The nature of the rocky substrate must be such that dished-out areas occur in the low intertidal region, and the exposure to wave action plays an important role as well. So, you won't find these habitats at every rocky area you visit, and of course, viewing them requires that the tide for the day be below the zero tidal level (a "minus tide," see Chapter 2). However, they are easily recognized for the unique group of organisms they support.

The Purple Sea Urchin. The first and foremost inhabitant is the purple sea urchin, *Strongylocentrotus purpuratus* (Color Plate 15a). The purple urchin is one of the most common marine animals in California. However, except for an occasionally stranded urchin tossed up by wave action or a hearty individual in an upper tidepool, it is often missed because it only occurs abundantly in the low intertidal zone and subtidally. This sea urchin cannot withstand long periods of exposure to warm air. Therefore, it inhabits the edge of the intertidal reef that is exposed only at extreme low tides, and low pools that are not exposed to the sun and air long enough for the water to heat up beyond the urchin's tolerance.

At very low tides (-1.0 foot MLLW or lower), large numbers of urchins can be found exposed on flat, seaward portions of the lower reef. In both this exposed, lowest intertidal region and in the low tidepools, the urchins occur in rounded depressions or pits (Figure 7-45). These pits are thought to be excavated by the

Figure 7-45. Solitary purple sea urchin, Strongylocentrotus purpuratus, *two inches in diameter, residing in its rocky excavation in a low intertidal pool.*

urchins using their spines and their special, five-jawed chewing apparatus called Aristotle's lantern. The urchin remains in its pit and feeds on any nearby attached fleshy algae or drifting plant material that is washed its way. As the urchin grows, it enlarges its pit, and if sufficient food is obtained, it will remain in place. The urchin sometimes becomes trapped in its pit, having grown too large to escape from the opening. The pit offers the round urchin a greater surface area for attachment of its anchoring tube feet, and protection from rolling rocks and other wave-borne materials.

In the low tidepool, the dark purple of the urchin is contrasted against the lighter pastel purple and pinks of coralline algae (Color Plate 2a and b). Corallines, both the encrusting and erect types, are very abundant here because their calcium carbonate-impregnated tissues are too hard for the urchins to graze. Thus, the pools are dominated by substantial erect coralline growth on the pool margins and encrusting corallines cover the bottom.

Low Tidepool Cnidarians. Other bright spots of color on the walls and floor of the low tidepool are provided by sponges of various shades of purple, red, yellow and off-white, and by cnidarians. Located in shaded corners or under overhanging ledges are dime-sized polyps of the orange cup coral, *Balanophyllia elegans* (Color Plate 5c). These are close relatives of coral reef corals and each forms a rigid calcium carbonate cup into which it can retract.

Another cnidarian that shows up here is the strawberry anemone, *Corynactis californica* (Color Plate 5d). Like the aggregating anemone, *Anthopleura elegantissima*, of the upper zones, the strawberry anemone is a clone-forming species. Also like the aggregating anemone, *Corynactis'* clones battle for space. Color Plate 5d shows two, different-colored clones in near proximity. Individual *Corynactis* polyps are about an inch across when open. The common colors are orange, pink, scarlet or red, but purple, brown, yellow, and nearly white clones are seen. The tentacles are tipped with white clubs. They are very common in the shallow subtidal as well.

The proliferating anemone, *Epiactis prolifera* (Color Plate 5b), is also common in the low tidepools, occurring in a range of vivid colors. Less varied in color, but nonetheless spectacular because of its size, is the giant green sea anemone, *Anthopleura xan-*

thogrammica (Figure 5a). Also clones and large, solitary individuals of the aggregating sea anemone, *Anthopleura elegantissima* (Color Plate 4d), inhabit these low pools.

Polychaetes. Two polychaetes are common here. On the bottom and edges of the pool, especially under overhangs, colonies of the filter-feeding worm, *Dodecaceria fewkesi*, form off-white patches (Figure 7-46). This polychaete secretes a calcium carbonate tube. The colony resides in a smooth white, matrix made up of these tubes. The colony is dotted with small (one tenth of an inch) openings through which the worm's head and tentacles protrude for filter-feeding. Twin spirals of red or pink tentacles about the size of a dime, protruding from white, calcareous tubes, mark the presence of *Serpula vermicularis* (Color Plate 7d). This filter-feeding fan worm was introduced previously in the under-rock section.

Low Tidepool Mollusks. Several mollusks contribute to these colorful tidepools. The lined chiton, *Tonicella lineata* (Color Plate 8c), is thought to feed on coralline algae and is frequently found in these low tidepools, sometimes in empty urchin pits. Another coralline-feeder found here is the dunce cap limpet, *Acmaea mitra*. Other gastropods occur in the low tidepools, including *Tegula* spp. and limpets. Occasionally, *Ceratostoma foliatum* and *Searlesia dira* (Figure 6-29) are seen as well. A common snail here is the small (half inch), but busy, *Amphissa versicolor* (Figure 7-47), seen moving actively about the surface of erect coralline algae. Often, this snail's shell has been appropriated by an equally active small hermit crab, so look closely.

Another striking snail is the beaded top snail *Calliostoma annulatum* (Color Plate 9b), with its half-inch high shell gloriously wound with purple beads. Sporting various gastropod shells are numerous hermit crabs, *Pagurus* spp. Occasionally, small clumps of mussels, *Mytilus californianus,* will appear in these low tidepools, probably dislodged from the rocks above. Another bivalve that shows up here is the rock scallop, *Hinnites giganteus* (Figure 7-21). It is distinguished by its bright orange mantle and blue eyes that are visible when the shell gapes open.

During the quiet, morning, minus low tides of spring and summer, the low tidepools feature the showiest of all the mollusks, the nudibranchs or sea

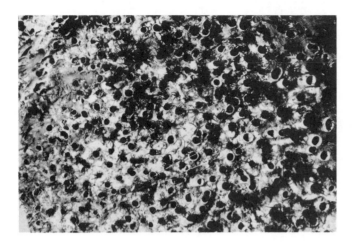

Figure 7-46. A colony of the colonial polychaete, Dodecaceria fewkesi. *Note the feeding tentacles projecting above the off-white tubes.*

Figure 7-47. The shells of the small snail, Amphissa versicolor, *commonly found in low tidepools, are frequently used by young hermit crabs.*

slugs (see Color Plates 10-12). Nudibranchs were described previously in the surfgrass section, but it is in the low tidepool, with its smaller area and beautiful background colors, that they can be most appreciated. Because the pools harbor sponges, cnidarians, and bryozoans that serve as food for many of the common nudibranchs, it is not unusual to find several species in a single pool. Look among the coralline algae and on the surface of the pool itself for nudibranchs crawling along upside down, using the surface tension created at the air-water interface as a foothold.

If there is any red sponge in the tidepool, look carefully for the camouflaged red sponge nudibranch, *Rostanga pulchra* (Color Plate 11c). *Rostanga* not only feeds on the red sponge, but also incorporates the sponge's pigment into its own body and that of its eggs that are attached to the sponge's surface. Look at the nudibranch photographed against a neutral surface (Color Plate 11d), and then try and find it on the host sponge (Color Plate 11c).

Sea Stars. Sea stars are seen in the low tidepools. The Pacific sea star, (*Pisaster ochraceus,* Color Plate 14b), sea bat, (*Patiria miniata,* Figure 6-28*),* leather star (*Dermasterias imbricata,* Figure 6-22), and the sunflower star, (*Pycnopodia helianthoides,* Color Plate 14a), are all occasionally discovered here, especially north of Point Conception. *Pycnopodia* is a voracious feeder on purple sea urchins and its mere presence in a low tidepool has been reported to cause the resident urchins to flee. Smaller species of sea

stars more commonly seen in these pools include small blood stars, *Henricia leviuscula* (Figure 7-17), and six-rayed sea stars, *Leptasterias* spp. (Figure 6-27). These two small stars often sport mottled colors that blend in with the coralline algae.

Low Tidepool Fish. In southern California it is not unusual to find juveniles (one to three inches long) of the fish known as the opaleye in low tidepools. Opaleye, *Girella nigricans* (Figure 7-48), as their name implies have vivid blue eyes. The body is olive-green and most are marked with one or two white spots on their backs, near the middle of their dorsal fins. This fish is a herbivore, and the adults are very abundant in kelp forests, feeding on the kelp plants.

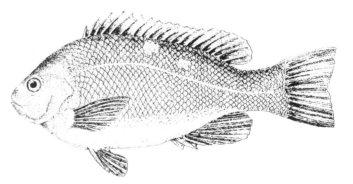

Figure 7-48. The opaleye, Girella nigricans, *a herbivorous fish of southern California kelp beds. Look for juveniles in low tidepools.*

Another, more rare, intertidal sighting in southern California is the juvenile Garibaldi, *Hypsypops rubicundus* (Color Plate 16b). The Garibaldi is California's state marine fish and is completely protected. Adults live in the shallow subtidal and are very territorial, challenging any intruder, including divers! Young Garibaldi differ from the bright, golden-orange adults. They are red-orange in color with the addition of iridescent blue spots and blotches along their flanks.

The other, urchin-dominated area of the reef, the low intertidal flat reef margin, is really somewhat different from the low tidepool, and might be considered a separate habitat. However, as this region receives considerable wave action and is exposed to air only at the lowest tides, it essentially supports a subset of the organisms described for the low tidepool.

Surge Channels

Physical Nature of the Surge Channel Environment. Surge channels (Figure 7-49) are formed by the differential weathering of the reef platform by the ocean. They are sometimes cut below the tidal level and thus never completely drain, even during the lowest tides. These submerged channels are typically at the very edge of the reef, and support large stands of the oar weed kelp, *Laminaria* spp. (Figure 7-50), and other kelps on the bottom, and feather-boa kelp, *Egre-*

gia menziesii (Figure 6-67), along the sides. Other surge channels extend well into the reef, in some cases reaching up into the mid-tidal level and above.

Besides tidal level, another variable in the surge channel habitat is orientation. Channels that extend directly into the reef, essentially perpendicular to the reef's edge, receive the direct force of the waves. Surge channels that turn to parallel to the reef's edge are quieter and receive a somewhat less forceful flow of water. Finally, the shape of the surge channels must be considered. Channels with straight sides tend to harbor a more meager cast of organisms than do channels that have substantial undercutting and overhangs. Channels with dished out bottoms tend to trap small boulders that bounce around and scour the walls, while channels with bottoms that slope continuously seaward are swept free of such material by wave action.

From this physical description you can see that the surge channel habitat is a varied one. What you discover in a given channel depends on all the variables listed and several others, including time of the year. One thing you should remember about surge channels: They are high-energy environments. They require that organisms be able to attach and hold on against the movement of strong water currents. They are also food-rich environments, in that the surging water contains many small organisms and organic materials swept from the reef and brought in from off-

Figure 7-49. Waves and a flooding tide enter the intertidal zone through a surge channel cut into an intertidal reef.

shore. It is not surprising then that many surge channel animals are attached filter-feeders that take advantage of this wave-borne bounty. A final note of caution, the same high energy that characterizes the surge channel can catch you unguarded! Watch out for waves. These are fascinating areas to explore, and you are more often than not bent over or on your hands and knees. Have a lookout watch for incoming swells and unannounced surges of chilling sea water. Remember, the surge channels are the avenues through which the tide floods into the intertidal zone. *Be alert.*

Surge Channel Overhangs. The organisms occurring along the top of the surge channel walls reflect the general organismal association for the particular tidal height and exposure. Thus, some walls are relatively barren, others are cloaked in thick algal growth, and still others support the spill-over from a well-developed mussel clump. However, it is in the shade of deeply undercut or overhanging walls in the lowest intertidal that the surge channel habitat reaches its zenith. (These strongly undercut niches are also found along the low intertidal, seaward edge of some semi-protected reefs). When you first observe one of these well-developed, low intertidal surge channel overhangs, you will be taken by the variety of colors, shapes, and textures.

Because the overhang is in deep shade, only the heartiest encrusting coralline algae will occur, and these are in competition with a variety of space-monopolizing, encrusting animal forms. The top of the roof of the overhang typically harbors the proliferating sea anemone, *Epiactis prolifera* (Color Plate 5b), and pale, individual aggregating anemones, *Anthopleura elegantissima* (Color Plates 4c and 4d). The heavily ribbed, reddish barnacle, *Tetraclita rubescens*, may also be found on the overhang roof. To many observers the oddest-appearing organism is the ostrich-plume hydroid, *Aglaophenia latirostris* (Figure 7-51), which hangs from the ceiling and sides of the overhang in feather-like colonies, up to three to four inches long. Other, more delicate hydroids can be discovered commonly in the surge channel overhang, but are beyond the scope of this guide and you are referred to more detailed sources [3, 4, 6].

Once you have identified a hydroid colony, take a close look at it to see an amazing, inch-long or smaller crustacean known as a skeleton shrimp (Figure 7-52). These animals are highly modified amphipods, called caprellid amphipods, and they co-occur with hydroids wherever you might find them. Compared to a more typical amphipod, caprellids have only a few appendages, and use them to grasp the hydroid colony firmly and to reach out to catch passing zooplankton prey.

Figure 7-50. The strong, flexible stipes of the low intertidal kelp known as oar weed, Laminaria *sp., hold the exposed plants upright during a very low tide.*

Figure 7-51. A colony of the ostrich-plume hydroid, Aglaophenia latirostris. *Each plume is two to three inches long and joined to the others by a common base.*

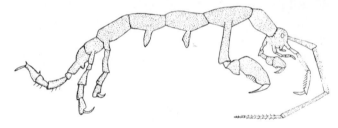

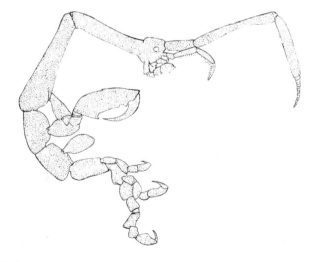

Figure 7-52. The amphipod crustaceans known as a skeleton shrimp, Caprella sp. *Look for these animals wherever hydroids are found.*

Sponges and Tunicates. The back wall of an overhang typically harbors several species of sponge growing in red, yellow, purple, and off-white sheets. Sponges are discerned by their soft, pliable texture and the numerous tiny holes on their surface. Another group of encrusting animals typically found here are compound tunicates, with sea pork, *Aplidium californicum* (Figure 7-53), being one of the most common. Like the sponges, the compound tunicates are filter-feeding animals. However, they are much more highly organized and are included in the same animal phylum, Phylum Chordata, as the backboned animals (vertebrates), which includes humans (see Chapter 3). Sea pork occurs in soft, mushy sheets that are typically off-white to light yellow or pink in color. Compared to sponges the compound tunicates have a slicker outer texture.

Hanging down from the roof of a surge channel overhang you may find the Monterey tunicate, *Styela montereyensis* (Figure 7-54). *Styela* is a solitary tunicate; it is not colonial. It has a yellow to dark red-brown, woody tunic and two obvious siphons for filter-feeding. *Styela* may reach ten inches in length and although they frequently have hydroids or other marine organisms growing on their tunics when found in quiet habitats, they are clean here in the wave-swept surge channel. Look here also for clusters of the light-bulb tunicate, *Clavelina hunstmani* (Figure

7-55). As the common name implies, *Clavelina* has a clear tunic that allows the bright-white elements of the animal's filter basket to show through like the filament of a light bulb. The clusters of light-bulb tunicates are asexually produced clones. The two-inch-long tunicates are attached at their bases.

Compound tunicates are the presumed prey of several small gastropod predators. Three of these are shown in Figure 7-56. The biology of these snails is not well known, and they are not especially obvious in life. However, the small, interestingly shaped shells seem to catch the beachcomber's eye, and they end up in many treasured shell collections as a result. The larger (to four fifths of an inch long), egg-shaped shell belongs to Solander's trivia, *Trivia solandri*. It is tan to brown in color with white ribbing and a distinct light furrow on the back. The aperture is narrow and borrowed by the ribs. It occurs from Los Angeles to Panama. The smaller egg-shaped snail (to two fifths inch long) is known as the coffee bean shell or the California trivia, *Trivia californiana*. It occurs from northern California to Baja. The shell is dark purple in color. The third species is known as the appleseed erato, *Erato vitellina*, and occurs from Bodega Bay in central California to Baja. The shell reaches three fifths of an inch in length, and is shiny and pear-shaped. The aperture is narrow and bordered by a light-colored area. The remainder of the shell is reddish.

Scattered among the other encrusting animals, colonies of the rosy pink bryozoan, *Eurystomella bilabiata* (Figure 7-6), can be seen with their typical, basket-weave design. Fan worms or featherduster worms occur here. The most common is typically *Serpula vermicularis* with its white, calcareous tube sometimes partially concealed in the burrow left by a boring clam. Another large fan worm that can be found here is the beautiful *Eudistylia polymorpha* (Figure 7-57), which lives in a five-inch long, parch-

Figure 7-53. An eight-inch-square colony of the compound tunicate, Aplidium californicum, *commonly known as sea pork. Note the colony's glossy appearance.*

Figure 7-54. Below: the Monterey sea squirt, Styela montereyensis; *above: a related species,* Styela clava, *found in quiet water habitats.*

Figure 7-55. A colony of the light bulb tunicate, Clavelina hunstmani. *Each sea squirt is connected to the colony via a common base.*

ment-like tube that hangs from the surge channel wall. The solitary orange cup coral, *Balanophyllia elegans* (Color Plate 5c), is also common here, typically along the base of the back wall.

Surge Channel Floor. Animals on the bottom of the surge channel overhang include more sea pork, sponges, and patches of the colonial polychaete, *Dodecaceria fewkesi* (Figure 7-46). The giant green sea anemone, *Anthopleura xanthogrammica* (Color Plate 5a), is frequently found on the bottom of surge channels. It waits here patiently for dislodged prey to be swept into its grasp. The principal food items for this anemone are sea mussels, *Mytilus californianus* (Figure 6-56), which are ripped loose from the clump by wave action. It must be a favorable feeding area for the anemone, as it reaches its largest intertidal size in the surge channels.

Occasionally, a large, red anemone will also show up in the surge channel. These are anemones in the genus *Tealia,* and four species occur in the intertidal zone of California, although they are more common subtidally. *Tealia lofotensis* (Color Plate 6a) is shown here, a truly beautiful animal.

Motile animals are also found in surge channels. Large cancer crabs, especially female *Cancer antennarius* (Figure 7-8) brooding eggs, seek out cracks and crevices along the walls. Several echinoderms appear to use surge channels as avenues in and out of

Figure 7-56. Three small, common low intertidal snails. Clockwise from the upper left: Trivia solandri, Trivia californiana, Erato vitellina.

Figure 7-57. The expanded feeding tentacles of a featherduster worm, Eudistylia polymorpha, *extend beyond its parchment-like tube.*

the intertidal zone, and may be encountered here on occasion. The Pacific sea star, *Pisaster ochraceus* (Color Plate 14b), and its two more subtidal cousins, *Pisaster brevispinus,* and *Pisaster giganteus (*Color Plate 14c), show this movement pattern. In central and northern California, the sunflower star, *Pycnopodia helianthoides* (Color Plate 14a), and the leather star, *Dermasterias imbricata* (Figure 6-22), may also move in and out this way, as does *Solaster stimpsoni (*Figure 7-18) in northern California. Six-rayed sea stars, *Leptasterias spp.* (Figure 6-27), blood stars, *Henricia leviuscula* (Figure 7-17), and sea bats, *Patiria miniata* (Figure 6-28), appear to be more per-

manent residents. Occasionally, a very large (up to six inches in diameter), red to scarlet urchin is found in a surge channel. This is *Strongylocentrotus franciscanus* (Color Plate 15b), the purple urchin's subtidal relative, which sometimes is carried into the intertidal zone by strong wave action.

Gastropods with tenacious grips occur in surge channels. *Tegula* spp. (Figures 6-15, 6-18, 6-19, 6-25), *Calliostoma annulatum* (Color Plate 9b), *Ceratostoma foliatum (*Figure 7-34), and abalones, *Haliotis* spp. (Figures 7-25 and 7-26) will be seen. Keyhole limpets will occur here occasionally, with the rough keyhole limpet, *Diadora aspera* (Figure 7-28), being the animal usually seen. Nudibranchs are here also. The sea lemon, *Anisodoris nobilis* (Color Plate 12a), the ring-spotted dorid, *Diaulula sandiegensis (*Color Plate 12b), and the red sponge nudibranch, *Rostanga pulchra* (Color Plate 11c and d), are among species that occur. Chitons are sometimes seen in surge channels with the vividly colored lined chiton, *Tonicella lineata* (Color Plate 8c), being the most common.

References

1. Behrens, D.W. *Pacific Coast Nudibranchs.* Monterey: Sea Challengers, 1991, 107 pp.
2. McDonald, G.R. and J.W. Nybakken. *Guide to the Nudibranchs of California.* Milbourne: American Malacologists. 1980, 72 pp.
3. Morris, R.H., D.P. Abbott and E.C. Haderlie. *Intertidal Invertebrates of California.* Stanford: Stanford University Press, 1980, 690 pp.
4. Smith, R.I. and J. Carlton. *Light's Manual: Intertidal Invertebrates of the Central California Coast,* 3rd ed. Berkeley: University of California Press, 1975, 716 pp.
5. Brusca, G.J. and R.C. Brusca. *A Naturalist's Seashore Guide-Common Marine Life of the Northern California Coast and Adjacent Shores.* Eureka (California): Mad River Press, 1978, 205 pp.
6. McConnaughey, B.H. and E. McConnaughey. *The Audubon Society Nature Guides. Pacific Coast.* New York: Alfred J. Knoph Inc., 1988, 603 pp.

8

Marine Mammals of California

❦

Over the course of the year, a total of 35 marine mammals occurs off the coast of California [1]. However, with luck you will probably only see a total of five or six. You may think it's pretty hard to miss a humpback or blue whale, but even though they do occur in California waters, they are normally found well offshore on the edge of the continental shelf. In this chapter I will introduce the most common marine mammals seen in California, and suggest where and when the beachcomber might find them.

Some very different groups of animals from the vertebrate class Mammalia have taken up life in the sea: the order Cetacea, the whales, including the suborder Mysteceti (baleen or whalebone whales), and suborder Odontoceti (toothed whales); the order Sirenia, including dugongs and manatees; and the order Carnivora, which includes the sea otters in the family Mustelidae, and the suborder Pinnipedia, including walruses, seals, sea lions, and elephant seals. As each of these groups adapted to the sea, they acquired a suite of similar adaptations. Their bodies became smooth and streamlined for ease of movement through the water, and the limbs became modified into flippers for swimming. All but the sea otters acquired a thick layer of fat or blubber for heat conservation and buoyancy.

Marine mammals are completely protected by the federal Marine Mammal Protection Act. It is illegal to take or possess any marine mammal or marine mammal part. Violation of this law carries a stiff penalty. In addition the sea otters are further protected by the Endangered Species Act. Beachcombers should appreciate these animals in their wild state. If you come across a dead or injured marine mammal notify the California Department of Fish and Game and they'll take it from there.

THE SEA OTTER

Researchers believe the sea otter, *Enhydra lutris* (Figure 8-1) once occurred continuously along the West Coast from Alaska into Baja California. In the eighteenth and nineteenth centuries American, Russian, French, and British fur traders eliminated the sea otter from most of its range. When a small colony was detected south of Monterey Bay off the Big Sur coast in 1938, people were caught by complete surprise because sea otters were thought to have been hunted to extinction everywhere on the West Coast except Alaska. From this original colony the sea otter has multiplied and spread north and south along the California coast. The otter now occurs from Ano Nuevo in Santa Cruz County all the way south to Purisma Point, just north of Point Conception in San Luis Obispo County. Over the last several years, the annual aerial census conducted by the California Department of Fish and Game suggests the otter population has stabilized between 1,500-2,000 animals.

Many biologists feel that the sea otter in California represents a unique threatened subspecies, *Enhydra lutris nereis*, and that it requires additional protection. In the 1980's the U.S. Fish and Wildlife Service (USFWS) transplanted a group of otters to San Nicolas Island of the California Channel Island chain, well away from the main coastal shipping lanes. The USFWS did this to create a buffer population that could be used to reintroduce the otter to the mainland coast in case it suffered mass mortality from an oil spill. We hope this will never be necessary, but after the carnage caused by the Gulf of Alaska spill, it is a sad contingency for which we must plan.

Sea otters closest relatives in the mammalian family Mustelidae are river otters. They are also related to

Figure 8-1. The sea otter, Enhydra lutris.

badgers, wolverines, weasels, skunks, and minks. The male sea otter reaches five feet long including the tail, and can weigh upwards of a hundred pounds. Females are smaller. The otter's large hind limbs are webbed and flipper like, and along with their flat tail, are used for swimming. Their forelimbs are short and stout, and their fingers stubby. However, otters are remarkable tool-users.

The sea otters' beautiful fur coat of fine, dense hair varies from reddish brown to black. Many older animals have silvery or white necks and faces. Because they lack a fat layer for insulation, the otter's coat is its only protection from the cold water. They spend much of their day grooming their fur as it looses its insulating capacity when soiled. This is why otters are so vulnerable to oil spills. Otters will often roll vigorously in the water to trap air bubbles in their fur for added insulation.

Sea otters can still move on land, although they rarely come ashore in California. In the ocean, otters spend most of their time on their backs when on the surface. They swim, eat, rest, carry their young, and sleep on their backs. Otters feed by diving to the bottom and searching for food. Their most active feeding periods are early morning and late afternoon. They feed in kelp forests and on sandy bottoms.

Sea otters eat anything that moves. The combination of a lack of a blubber layer for insulation and the cold Pacific water places a tremendous energy demand on them. A mature otter must eat an average of 25% of its body weight daily to maintain itself. It accomplishes this by eating the biggest, most energy-rich organisms it can catch. Abalones and sea urchins are the preferred prey when an otter first enters a kelp forest feeding area. The otter is known to use a rock to break the abalone's shell so it can be pulled off its rocky perch. Otters will also swim on their backs, cradling a rock on their chest that they use to crack open sea urchins. Otters will fairly quickly eliminate these large prey in a new feeding area. In Monterey Bay the only sea urchins or abalones you see underwater are in deep crevices, well out of the otter's reach. Many feel that the otter's fondness for sea urchins goes a long way towards maintaining the health and vigor of the kelp forest. Sea urchins can reach high population densities and literally strip a kelp forest of vegetation. In the face of the otter's continued predation, sea urchins will always be kept in check.

Once established in an area, the otter's diet mainly includes large cancer crabs, shellfish, and whatever fish they can catch. Otters have been seen eating sea stars by biting off the tip off the arm and sucking out the innards! Sea otters are also very effective at feeding on sand bottoms, taking clams and crabs.

The California sea otter is truly a marine mammal. They mate in the water, and the female gives birth there as well. Female sea otters bear a pup about every one to two years between February and June. The female is a devoted mother, carrying her pup with her as she swims on the surface, and leaving it temporarily in the kelp as she dives after food. They are active during the day and the normal behavior at night is to wrap themselves in the surface kelp canopy and sleep. The otters' chief natural predator appears to be the great white shark.

Viewing Sea Otters. The best place to see sea otters is Monterey Bay. These waters are relatively protected with extensive kelp forests, and the otters are year-round residents. They often ply the kelp forests immediately offshore, and can be viewed from the shore with binoculars. Occasionally, the otters are close enough to be seen with the naked eye, especially when they come nearby the wharves and the viewing areas of the Monterey Bay Aquarium in Pacific Grove. Look for them during their active feeding periods in the early morning and late afternoon. Watch for sea gulls hovering over the kelp, waiting for the otters to surface so they can pick off the scraps of their meal.

Friends of the Sea Otter. The concern for the sea otters' continued success and safety along the precarious 220-mile stretch of California's coast they inhabit has led to the formation of a non-profit citizens' group call the Friends of the Sea Otter. Beachcombers wishing more information about these amazing mammals should contact this group at their Carmel headquarters: Friends of the Sea Otter, P.O. Box 221220, Carmel, California 93922; phone 408-625-3290.

PINNIPEDS

The Latin term "pinniped" literally means "feather-foot," and is a reference to the seal and sea lions' flippers. Pinnipeds are thought to be most closely related to bears, and anyone who has gotten too close to a mature male sea lion will certainly attest to the similarity! There are two main types of pinnipeds, the family Phocidae and the family Otariidae [2]. Otariid pinnipeds are also known as the eared seals, because they have visible ear flaps on the sides of their heads. Otariids include the sea lions and the fur seals. These animals swim with their large fore flippers, and their rear flippers can rotate under the body to support it on land. Otariids are quite agile on land and can move quite rapidly. The traditional "trained seal" of circus and marine animal parks is most likely a female California sea lion. In addition to the California sea lion, the Stellar sea lion is also commonly seen in California. The northern fur seal is also here part of the year, but remains offshore and will not be discussed in this guide.

Phocid pinnipeds lack a visible ear flap, and have a simple ear opening on the sides of their heads. They swim by sculling with their rear flippers. They usually hold their fore flippers close to their bodies and use them to turn and maneuver in the water. The rear flippers point backward and cannot be rotated forward under the body to support the phocid on land. Their fore flippers are not positioned to hold the upper body erect. Consequently, these pinnipeds are much less adept at moving across open ground on land, and scoot and bounce along on their bellies inch-worm fashion. Two phocid pinnipeds are relatively common in California, the harbor or leopard seal, and the northern elephant seal.

The Harbor Seal

I'll start with the harbor or leopard seal, *Phoca vitulina* (Figure 8-2), because it is found year-round in California. The harbor seal is California's smallest pinniped. Males may reach five to six feet in length and weigh up to two hundred pounds. The females are somewhat smaller. The seals are silver-gray colored with dark brown to black spotting, which gives them the "leopard" common name. Some individuals are very dark, almost black, and the spots barely show through. Harbor seals have a broad geographic range. They are found from the Arctic to Baja California in the east Pacific, occur widely in the west Pacific, and on both sides of the Atlantic as well. Some experts recognize a unique subspecies, the Pacific harbor seal, *Phoca vitulina richardsii,* found only along the west coast of North America.

As the common name harbor seal implies, these animals can be found nearby humans. They haul out (come on shore to rest or sleep) on sand bars in river mouths and estuaries, and on rocky intertidal reefs at low tide. Harbor seals do not occur in large aggregations like the other pinnipeds. Usually six or a dozen at most will be seen, and occasionally up to three hundred may gather on a large sand bar. The California population is estimated to be somewhere around 17,000 [3], with the majority found north of Point Conception.

Figure 8-2. Four harbor seals, Phoca vitulina, *hauled out on a middle intertidal rock. Note the rear-facing hind flippers and small fore flippers.*

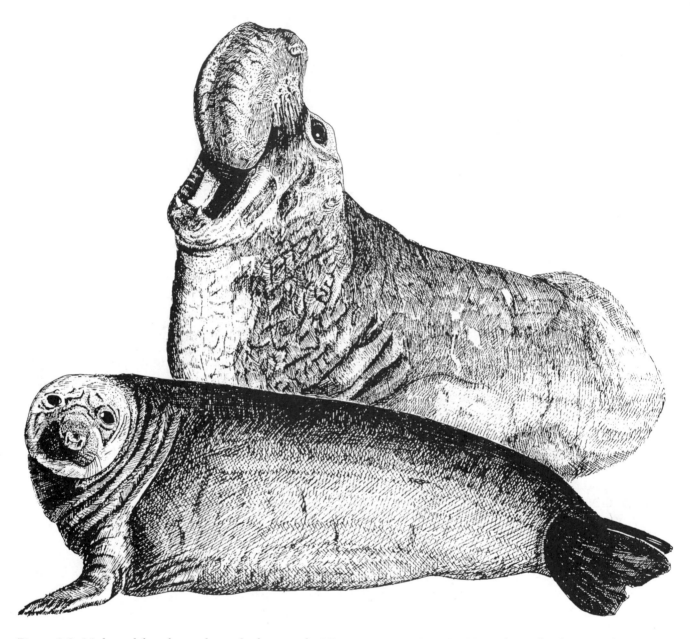

Figure 8-3. Male and female northern elephant seals, Mirounga angustirostris. *Note the male's large proboscis.*

On land harbor seals are very skittish and will quickly return to the water when approached. However, in the water they are often quite curious. When quiet in the water, harbor seals float vertically with just their head above water. They don't dive to submerge like other pinnipeds, instead they just sink and swim off horizontally. They also often swim with their head above water. It is not unusual to have a harbor seal follow you along a sandy beach, swimming at a safe distance. I have often had a curious harbor seal watch and follow my dog as we walk along the beach near our home. My most memorable experiences with harbor seals have occurred while scuba diving in Monterey Bay. Harbor seals frequently swim down and look over my shoulder while I work on the bottom. On several occasions I have also involuntarily "shared" fish I've speared for class with an aggressive harbor seal. The surprising thing about those encounters was I never felt threatened by the seal. It always seemed like a sporting proposition on both sides.

When they are not helping themselves to divers' catches, harbor seals catch their own fish, crabs, and cephalopods (squid and octopus) as their main diet. They eat primarily non-commercial fish species in shallow water, so they are not considered a nuisance by fishermen like some of their pinniped cousins.

Harbor seals mate promiscuously, that is they don't form permanent pair bonds. They also don't come together in big mating aggregations and end up with a few males dominating the group like the other pinnipeds. Harbor seals give birth on land from May to July, and the pup is an adorable creature. The pups learn to swim fairly early, but initially are not strong enough swimmers to accompany their mother on feeding forays. The female will leave her pup unattended on a beach or rocky reef while she forages for food. This leads to an unfortunate scenario of concerned humans "rescuing" a supposedly abandoned pup from the beach. Meanwhile, the mother watches frantically from the water, too frightened to come ashore. *Leave harbor seal pups alone.* If they really appear in distress and no adult is in sight, or if they are entangled in something, call the California Department of Fish and Game or the local animal shelter, don't attempt to take the pup in yourself.

Northern Elephant Seal

The second phocid seen in California is the northern elephant seal, *Mirounga angustirostris* (Figure 8-3). As can be seen from the illustration, this is not an animal you could cuddle up to. Males reach 16 feet in length and weigh up to 5,000 pounds! Females are considerably smaller, reaching over 9 feet and 2,000 pounds. The animals are brown to buff-colored when dry, and the sexes look very different. Females are very seal-like with big cow eyes, while the males always remind me of Jimmy Durante. Upon reaching sexual maturity at about 9 years of age, the male's snout becomes greatly enlarged into what is called a proboscis. The proboscis is inflatable, and serves as a resonating chamber for the male's deep-throated vocalizations made during mating season.

Elephant seals are huge animals. They feed far out to sea and at incredible depths. Their prey includes squid and fish. Each year the adults reassemble on the rookery beach of their birth for the annual mating season. In California the elephant seal rookery beaches are on the Farallon and Channel Islands, and at Ano Nuevo in southern San Mateo County. The males arrive first on the rookery in December and go through prolonged battles, both mock and real, to establish a hierarchy. As can be seen in the illustration, the bull elephant seal can raise his upper body upright. Bulls face each other and make slashing attacks to each other's chest area (Color Plate 16c). They have an especially thick layer of skin in the chest and lower neck area, but it still can be a very bloody affair. The top males in the hierarchy are called beachmasters or alpha bulls. These animals don't defend specific territories like the sea lions, but instead position themselves on the rookery where the most females congregate.

Females do not arrive at the rookery all at once, but over a two-to-three-month period. Shortly after their arrival they give birth to a 60-pound pup. The female becomes sexually receptive a few days after her pup's birth and mates with the alpha bull. She will nurse the pup for four weeks. When she is through, the pup will weigh up to 300 or more pounds! After weaning her pup the female leaves the rookery to return to the sea and replenish herself.

During the non-breeding season the elephant seals return to the rookery to go through an annual molt. They actually shed the outer thick layer of skin along with their hair, a process that takes three to four weeks. The males, females, and sexually immature animals come in to molt during different periods of the spring and summer.

Viewing Elephant Seals. The entire shore-based part of the elephant seal's life can be viewed by California beachcombers. As I mentioned, Ano Nuevo State Park in southern San Mateo County is the site of a major elephant seal rookery. Each year thousands of people are led by docents out across an active dune field to view the activities on the rookery beach. During the non-breeding season, people can go out on their own to see the animals that have come onshore to molt. Ano Nuevo is a veritable marine mammal paradise! Stellar sea lions use Ano Nuevo Island, visible across a narrow isthmus, as a rookery during the summer, and a large contingent of resident harbor seals is also usually at hand.

Elephant seals are a remarkable success story. Like the sea otters they were hunted to near extinction by marine mammal hunters. Elephant seals were hunted for their oil, not their coarse, thick hides. In the 1910's they were reduced to an estimated [1] population of fewer than a hundred animals on Guadalupe Island located off the west coast of Baja California. Once given protection by both the United States and Mexico, they have come back with a vengeance. They now number over 65,000 [3], and their population continues to grow. Beaches that are known to be the locations of historic rookeries are being re-established on offshore islands and remote mainland California sites.

California Sea Lion

There is a beautiful place on the central California coast in Monterey County known as Point Lobos. Lobos means wolves in Spanish; however, the namesake wolves in question here were probably vocalizing California sea lions, *Zalophus californianus* (Figure 8-4). The barking of California sea lions is unforgettable once you've heard it. For many these animals somehow characterize the happy-go-lucky, fun-loving image that Californians are noted for. After watching them cavort around the rocks, body surf choice waves, and steal salmon off anglers'

hooks, it looks like the good life. However, it isn't all sun and siestas. Take a close look at the photograph of the sea lions. You'll notice an unmistakable wound on the right flank of the upper animal. This may have been caused by a large predatory shark or killer whale, or a dispute with a commercial fisherman. California sea lions live in a real and often hostile world.

The California sea lion is the most abundant and most commonly seen pinniped along the California coast. Mature males average seven feet or more in length. They usually weigh between 500 to 750 pounds, with an occasional thousand pounder observed. Adult males are readily distinguished by a thick neck and a pronounced ridge in the center of the head called a saggital crest (Figure 8-5). The female lacks the crest and may reach 6 feet in length and 250 pounds. The animals appear black when wet, with the males drying to a dark brown and the females usually a lighter brown. The flippers are hairless and black in color.

The California sea lion is found all along the California coast. Traditionally, they haul out on sandy beaches and flat reefs of offshore islands, and at relatively inaccessible sites on the mainland. However, during the late 1980's several animals found the pickings from the tourists so rich that they elected to hang around Monterey at Fisherman's Wharf and the Coast Guard breakwater. Another batch has taken over Pier

Figure 8-4. Two California sea lions, Zalophus californianus, *sun themselves. Note the wound in the flank of the upper animal.*

Figure 8-5. An adult male California sea lion readily distinguished by his thick neck and the pronounced ridge in the center of his head.

39 in San Francisco, and driven the boat owners from their moorings. They have become quite a tourist attraction, but stay upwind.

Male California sea lions move south to join the females on their chief California rookery sites on San Miguel and San Nicholas Islands of the California Channel Island group. Similar to the elephant seals, a few dominant males monopolize the breeding that occurs from June through July. The female gives birth to a single pup and may nurse it for several months up to a year. After the breeding season the males may travel as far north as British Columbia with the females staying more to the south near the rookeries. You may encounter California sea lions mixed in with other pinniped species. In areas where they overlap, it is not unusual to find California sea lions sound asleep on top of sleeping elephant seals. California's pinnipeds are all very gregarious during their non-breeding seasons (Color Plate 16d), and often haul out in mixed species groups.

Sea lions are exceptionally strong, graceful swimmers. With their long fore flippers they virtually fly through the water. When in a hurry they sometimes "porpoise," leaping out of the water. They have been estimated to swim at speeds of up to 25 miles an hour. Sea lions eat mainly fish and squid. They have a very bad reputation with commercial fishermen for stealing fish from nets and off hooks. However, studies on their diet reveal they consume mostly non-commercial fish species.

Stellar Sea Lion

The Stellar sea lion, *Eumetopias jubata* (Figure 8-6), is the largest of our eared seals. The males average nine and a half feet in length and weigh up to two thousand pounds. Females average six to seven feet and six hundred pounds. Stellar sea lions are light yellowish in color, varying from almost cream color to yellowish brown, with black flippers. These animals are distinguished from the California sea lions by their larger size, lighter color, and the lack of the incessant, staccato-like bark of the California sea lion. The male stellar lacks the pronounced crest of the California, and has long, coarse hairs on his neck that resemble a mane. This mane, along with the Stellars' penchant for roaring at one another while defending territories on their breeding grounds, may be the sources for the common name "sea lion."

Stellar sea lions typically don't enter harbors or estuaries like the more adventuresome Californias. They are animals of the open coast, preferring offshore rocky reefs as haul-out sites. A good place to view Stellar sea lions is Seal Rock below the historic Cliff House on the San Francisco coast.

Stellar sea lions can occur all along the California coast, but are more common north of Point Conception. Their chief California breeding rookery is Ano Nuevo Island, where the males begin to arrive in May. Like the California sea lions, an individual male may hold a harem of up to 30 or more cows, although

Figure 8-6. Male and female Stellar sea lions, Eumetopias jubata.

it is usually smaller. The females arrive in June, give birth to a single pup, mate, and are gone by the late summer. Stellar sea lions head north after the breeding season. Aggregations of animals may be viewed on offshore rocks in northern California during the winter, but most of the population is out of the state. Males go to the rich feeding grounds in the Bering Sea to feed.

The Stellar sea lion population has been in decline in the last two decades. One reason given is the stiff competition the animals are facing for their once-plentiful food on their northern winter feeding grounds. Stellars feed mainly on fishes not used as food in the United States. However, foreign fishing fleets in cooperation with the United States, have been taking wholesale amounts of these fish, and the loser appears to be the Stellar sea lion.

THE WHALES

Whales are divided into two categories, the baleen whales (suborder Mysticeti), and the toothed whales (suborder Odontoceti). The baleen whales lack teeth, and instead have horny, fibrous structures called baleen plates lining their upper jaws. These plates overlap one another and are smooth on their outer surface, while the fibrous bristles on the inner surface are frayed and cover the space between the plates. They feed by engulfing great quantities of water containing prey species. They force the water out between the plates with their enormous tongues, and trap their quarry on the frayed bristles on the inner side of the baleen plates. Prey varies from small fishes and squid in some species to shrimp-like krill and other zooplankton. Baleen whales include the large, majestic whales like the blue and sei whales, the singing humpback whale, and the gray whale, which I will cover in this chapter.

Toothed whales include the sperm whale, killer whale, and the dolphins and porpoises among others. As the name implies, these mammals have a jaw full of conical, pointed teeth. They feed by capturing fish and other large marine animals. On any given day a beachcomber may encounter the bottle-nosed dolphin (*Tursiops truncatus*) body surfing along southern California beaches, and Pacific white-sided dolphins (*Lagenorhynchus obliquidens*) often ride the bow wave of larger vessels offshore. I don't have room to cover all these delightful animals, and recommend

you consult the references at the end of the chapter [3, 4, 5] for more information. I will discuss only one toothed whale that the beachcomber can't miss if it is around, the killer whale.

The Gray Whale

Californians are very proud of their gray whales. We call them California gray whales, although nobody else does. The fact is they don't belong to California at all, but to the entire eastern north Pacific. Gray whales undertake one of the longest annual migrations known for any mammal, a round-trip journey of over 13,000 miles. Each year they move from their winter breeding grounds in the coastal lagoons of Baja California and mainland Mexico to their feeding grounds in the Bering Sea north to the edge of the Arctic ice cap. During this migration the gray whales remain close to shore, and are a frequent sight all along California during their November to May migration period.

The gray whale, *Eschrichtius robustus* (Figure 8-7), is a medium-sized whale, ranging from 35-50 feet long and weighing up to 35 tons. The females tend to be slightly larger than the males for a given age. The animal is mottled gray with extensive white streaking, and has light-colored patches of barnacles all over its body, especially the back, head and lower jaw. Gray whales lack a dorsal fin and instead have a midline ridge that extends from the head to a series of bumps or "knuckles" near the tail.

The life cycle of the gray whale is timed around the annual migration. Females give birth to a single calf in the warm breeding lagoons between January and March. The interval between calves is at least two years. The calf is born tail-first underwater and must surface to take its first breath. Gray whale calves are about 16 feet long at birth and weigh about a ton. The mother nurses them with a milk that is richest known of any mammal, containing 40% fat and 40% protein with very little sugar. Females with calves don't begin their northern migration until late March, eight weeks after the rest of the pack. They are visible along the California coast from April to June.

The gray whales spend from May to November feeding in the Bering, Chukchi, and Beaufort Seas. The feeding of the gray whale is unique among all the baleen whales. They eat almost entirely benthic amphipods that they obtain by stirring up the muddy

Figure 8-7. A female gray whale, Eschrichtius robustus, *and her calf.*

bottom. They move their feeding northward as the ice pack melts in summer, and are driven southward as it reforms in the fall, finally relenting and beginning their fall migration.

Gray whales don't migrate in large groups. As many as 20 have been sighted traveling together, but singles, duos and trios are far more common. As the whales move along the coast they can be observed jumping out of the water or breaching, or holding their heads vertically out of the water, a behavior known has "spyhopping." Some believe the whales spyhop to look for familiar coastal landmarks to guide them in their migration.

Gray whales will occasionally come into estuaries along their migration route. Almost every year two or three cruise into San Francisco Bay, check out the action at Fisherman's Wharf, and then swim back out to continue their journey. Gray whales also frequently come into very shallow water along the open coast, especially near sandy beaches. Along the beaches they are sometimes seen in the surf line, and researchers believe they may be attempting to scrape the barnacles off their skins.

Viewing Gray Whales. The best time for California beachcombers to view the southern gray whale migration is in December and January. Migrating whales can be seen from shore at a number of sites in California. Some of the most popular are Point Reyes in Marin County, Davenport Landing in northern Santa Cruz County, Point Lobos and Big Sur in Monterey County, Dana Point in Orange County, and Point Loma in San Diego County. In addition whole and half day, whale-watching cruises are becoming widely available out of California's coastal ports. These can be hard on people prone to seasickness or very small children, so be mindful of the ocean conditions before such a venture.

A final word on gray whales. They join the sea otters and elephant seals on the list of marine mammal survivors. There were once populations of gray whales in the Atlantic and in the west Pacific. These were hunted to extinction. The east Pacific population was likewise in trouble when the gray whales were protected by international treaty in 1946. Since then the gray whale has bounced back, and is now estimated at a population size of about 15,000-21,000 [2,

Figure 8-8. Two killer whales, Orcinas orca. *Note the large dorsal fin of the male in the foreground.*

3]. In fact the gray whale's recovery has been so solid that it was removed from the endangered species list in 1993. Gray whales are still taken for food in small numbers on their northern feeding grounds by indigenous tribes. However, their chief predator is thought to be the killer whale, especially on young whales during the migration.

The Killer Whale

There is no mistaking the killer whale, *Orcinas orca* (Figure 8-8), for any other marine mammal. Its striking black and white coloration, large, rounded, flippers, and tall dorsal fin make it easily recognizable even far offshore. Once feared and loathed for their predatory activities, the performing orcas of marine theme parks have completely changed the public's perception of these magnificent animals. Mature male orcas may reach 30 feet long and weigh 8 tons or more. Females are smaller, reaching 23 feet and 4 tons. Killer whales have a small white patch immediately above and just behind the eye, and a large white ventral (bottom) patch that extends from their chins to their tails and extends up on either side of their rear flanks. They also have a gray saddle just behind the dorsal fin.

The dorsal fin of the female and immature males is curved (falcate). At sexual maturity the male dorsal fin becomes steeply triangular, reaching six feet in height. The shape, pigmentation, and scaring of the dorsal fins and tail flukes are often unique, and have allowed researchers to identify and trace individuals over many years. From research done on killer whales in the Pacific northwest and elsewhere, quite a bit is known about these most successful hunters.

Killer whales are often likened to wolves [3]. They live in stable groups with complex social orders, hunt mammalian prey cooperatively, and are top predators. Killer whale groups are called pods, and their make-up consists of adults of both sexes and immatures. In the Pacific Northwest these pods are year-round residents in fairly distinct areas. These pods tend to feed mainly on fish, chiefly salmon.

Other pods are known to be more nomadic. Killer whales are known in all seas at all latitudes. Seasonal movements of pods seem to be tied to ice conditions in polar areas, and the availability of food in other areas [3]. These pods feed primarily on other marine mammals, including seals, dolphins, and even the largest of all animals, the blue whale. When attacking larger prey like gray whales, the killer whales have been observed to herd and attack their quarry in a very deliberate, coordinated fashion. They often take only the tongue of these large whales, leaving the rest for sharks. There is virtually no large marine organism that is safe from the killer whale.

One of the favorite haunts of these nomadic pods is the seasonally occupied rookery of pinnipeds. They are also seen near feeding areas or haul out areas of pinnipeds, like the mouths of rivers or estuaries. The California beachcomber wishing to view the killer whale somewhere other than a marine theme park should watch the local newspapers and news telecasts for reports of sightings. Usually, the pod will remain in an area for some time, before moving on to the next food source.

References

1. California Coastal Commission. *California Coastal Resource Guide*. Berkeley: University of California Press, 1987, 384 pp.
2. Riedman, M. *The Pinnipeds, Seals, Sea Lions, and Walruses*. Berkeley: University of California Press, 1990, 439 pp.
3. Leatherwood, S. and R.R. Reeves. *The Sierra Club Handbook of Whales and Dolphins*. San Francisco: Sierra Club Books, 1983, 302 pp.
4. Orr, R.T., and R.C. Helm. *Marine Mammals of California*. Berkeley: University of California Press, 1989, 92 pp.
5. Daugherty, A.E. *Marine Mammals of California*. Sacramento: California Department of Fish and Game, 1972, 90 pp.

Index